I0468190

God's Own
The Genesis of Mathematical
Story-Telling

Nma (Beautiful) Jacob

God's Own: The Genesis of Mathematical Story-Telling
By
Nma (Beautiful) Jacob

Disclaimer

All the material contained in this book is provided for educational and informational purposes only. No responsibility can be taken for any results or outcomes resulting from the use of this material. While every attempt has been made to provide information that is both accurate and effective, the author does not assume any responsibility for the accuracy or use/misuse of this information.

First Edition 2015

Published in the United Kingdom by
GGEC UK

For Publisher inquiries
GGECUK
Empowerment House
London
www.joshuajogo.com
ggecuk@gmail.com
+447448522120

Presented To:

_____

From:

_____

Date:

_____

Sign:

_____

God's Own: The Genesis of Mathematical Story-Telling

"I am a NiWARD Talking Drum, Shekere Musical Instrument
and Akwete Cloth!"

A NiWARD Autobiography of
<u>Nkechi Madonna Adeleine Agwu, Ph.D.</u>

2014 Carnegie African Diaspora Fellow and
Mathematics Story-Teller of Members of the Organization of
Nigerian Women in Agricultural Research for Development (NiWARD)
Centre of Gender Issues in Science and Technology (CEGIST)
Federal University of Technology, Akure (FUTA), Nigeria

Professor of Mathematics and
Past Director of the Center for Excellence
in Teaching, Learning and Scholarship
Borough of Manhattan Community College (BMCC)
City University of New York (CUNY), United States of America

Table of Contents

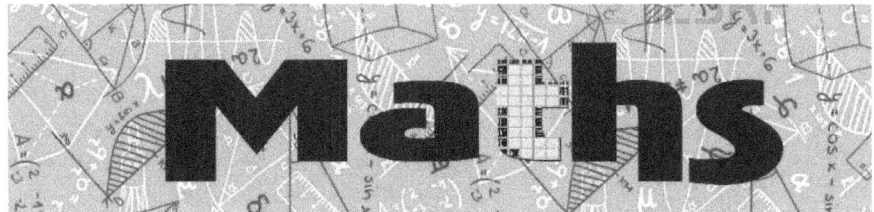

Dedication

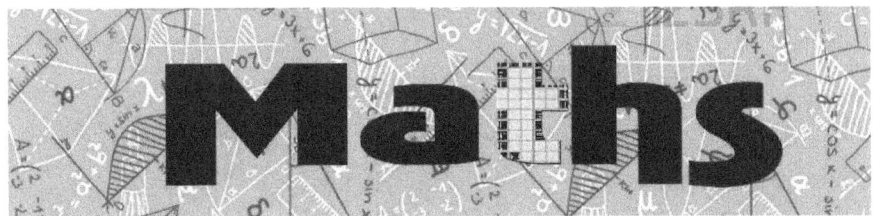

First and foremost this book is dedicated to *The Absolute Infinite* (God), the author and finisher of our faith, the Alpha and Omega, who guided my pen in writing my NiWARD autobiography. Imela Okaka Onyekeruwa (Thank you Great and Mighty Creator of the World).

It is dedicated to my late father, Mr. Jacob Ukeje Agwu, a 9th generation farmer from Umuapu, Agbakoli, Akoliufu, Alayi. He taught me the basics of scientific investigation. He planted in me the spirit of creativity, innovation and adaptation. From him I adopted the philosophy of "do what you can, from where you are, with what you have."

It is dedicated to my late paternal grandmother, Mrs. Omamma Virginia Agwu, nee Egu Agu. She was a humble market woman and 8th generation farmer from Umuapu, Agbakoli, Akoliufu, Alayi. She was an expert Okwe (Mancala) game player. She taught me how to play this game. This initiated my research journey and scholarship in African culture, and women's stories in science, technology, engineering, and mathematics (STEM) related fields.

It is dedicated to my mother, Mrs. Europa Lauretta Durosimi Wilson-Agwu. She is my 1st female mentor. Her *Red Cross* volunteer work caring for and transporting critically ill Biafran children from Biafra to the safety of refugee camps in Gabon and Fernando Po (now Equatorial Guinea or Bioko) inspires me.

Finally, it is dedicated to my beloved son, Mr. Ngozichukwuka Jacob Ayemere Durosimi Agwu, in whom I am well pleased. He gave me a present in 1998 by surviving distress in labor on my birthday. According to his name, Ayemere (Immortalize My Name) I am expecting him to tell the story of Nma (Beautiful) Jacob to his children and to the world when I depart from this earth.

Acknowledgements

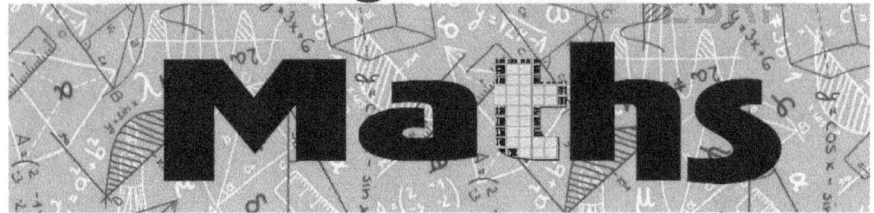

I would like to acknowledge the following people who inspired, motivated, and facilitated the writing of my NiWARD autobiography - Professor Emeritus Stella Williams, Professor Bronislaw Czarnocha, Ms. Iman Drammeh, Mr. Jeremy Coats, Elder Austin Tuitt, Comrade David Kalu, Prophet (Rev.) Emmanuel Angel, Councilman (Apostle) Joshua Jogo, and three of my student mentees at the Borough of Manhattan Community College, City University of New York (BMCC CUNY) – Ms. Saduda Oyo, Mr. John Aderounmu, and Mr. Michael Villalona. I would like to acknowledge the following organizations that facilitated the writing of my NiWARD autobiography – Carnegie Foundation; Institute of International Education (IIE); BMCC CUNY; Centre for Gender Issues in Science and Technology, Federal University of Technology, Akure (CEGIST FUTA); Nigerian Women in Agricultural Research and Development (NiWARD); Mathematics Teaching-Research Journal (MTRJ); Drammeh Institute; Pan-African Strategy and Policy Research Group (PANAFSTRAG); Nigerian Mathematical Centre (NMC); African Views Organization; Bende Analyst Newspaper, and United Peoples Inclusive (UPI).

Lastly, I would like to thank all my colleagues, friends and mentors, namely, Rev. Leo Woodberry, Bishop (Dr.) Joe Omeokwe, Bishop Ebony Kirkland, Dr. Nemata Majeks-Walker, Elder Austin Tuitt, Venerable Emeka Nwigwe, Rev. Fada (Dr.) Jon Ukaegbu, and Prophet (Rev.) Emmanuel Angel who wrote the foreword and friendly reviews to accompany my NiWARD autobiography, and others who will be carrying out subsequent reviews. I would also like to thank the members of my family (immediate and extended), all the great women and men of God who are members of Vineyard International Christian Ministries and of Glorious Miracle Embassy International, and all others, for their intercessory prayers on my behalf professionally, physically, mentally, financially, and otherwise, during the period of incubation and writing of this autobiography. I could not have reached the finish line without your intercessory prayers. Imela! Imela! (Thank you! Thank you!).

Foreword

*G*od's Own: The Genesis of Mathematical Story-Telling is the autobiography of my friend Dr. Nkechi Madonna Adeleine Agwu. It is more than just a single story about a very brave and successful woman. Like the woven piece of Akwete cloth she uses to teach mathematics, her story is intricate. It is made up of the many fibers who are her parents, grandparents, child, trials, tribulations, struggles, and successes that marked her life as unique.

Dr. Agwu's life began as a child of a Nigerian farmer. Her father and mother both valued learning and education especially in the fields of mathematics and science. That is where her academic journey began. Yet her journey was not one traversed on the ivy covered pathways of academia alone. As a child barely five years old she saw the real world. At a time when she should have lived in a world of children's fables, fairy tales, and sugar plums, she learned the brutal vicissitudes of war. Her family became the victims of the bitter Biafran or Nigerian Civil War, where ethnic groups where pitted against one another. People, young and old were deliberately starved to death or battered into eternal rest by bullets and bombs. Over two million people died. It was an African holocaust that is but should not be forgotten. Forced from their homes the Agwu family lived both as homeless, displaced or refugees, depending on your predilection for political correctness. That did not stop the future doctor. Instead of becoming callous by her experiences she became more compassionate, more Christ-Like. Neither did she forget other young girls like herself. She yearned for an education and an opportunity to be molded not by the world but as God's own creation filled with the spirit of our Maker's creativity.

In an ever shrinking and competitive world where science, technology, engineering and mathematics (STEM) holds the keys to the future, Dr. Nkechi Madonna Adeleine Agwu has developed the tool of "mathematical story-telling" to equip new generations so that Africans and others worldwide can rise above social, political, economic and gender inequality, and injustice and soar in a brand new world. Bending not to the burden of conformity of suffering but standing to embrace the power of transformation, such is the being of Nkechi (God's Own). Thank you, my friend for helping to show us one of the ways.

And be not conformed to this world: but be ye transformed by the renewing of your mind, that ye may prove what is that good, and acceptable, and perfect, will of God. Romans 12:2.

Rev. Leo Woodberry
Executive Director of Woodberry & Associates
Pastor of Kingdom Living Temple
Telephone: +1- 843 410 3272
Cell: +1-843 410 3272
www.WNAConsulting.net
www.KingdomLivingTemple.snack.ws

Introduction

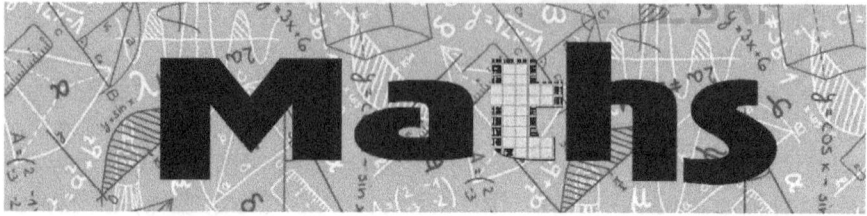

In the beginning God created woman in His own like image, blessed her to go out to be fruitful and multiply, to replenish the earth, subdue it, and have dominion over everything on it, Genesis 1: 27-28. *God's Own: The Genesis of Mathematical Story-Telling* is about the creativity of God (The Absolute Infinite) as He is reflected in women. It is about the power of The Absolute Infinite to perform signs and wonders reflected in mathematical symbols, patterns, numbers, and graphs. It is a book on His word according to John 1:1, in the beginning was the word, the word was with God, and the word was God.

God's Own: The Genesis of Mathematical Story-Telling is a mathematical, cultural, and spiritual story of Dr. Nkechi (God's own) Madonna Adeleine Agwu, aka Nma (Beautiful) Jacob. Dr. Agwu is an African female leader in mathematics and statistics education. Her life trajectory includes experiences of displacement, homelessness, living in refugee camps, single-parenting of a child with hearing and speech needs, and many other issues that from all indications could have set her up for failure, but for God's divine intervention.

God's Own: The Genesis of Mathematical Story-Telling is about Dr. Agwu's journey as an ethno-mathematician to her present scholarship in African culture, and women's stories in science, technology, engineering, and mathematics (STEM) related fields. It is about how she found her identity as a mathematical story-teller of members of the Organization of Nigerian Women in Agricultural Research for Development (NiWARD). It is a story about the beauty, strength, courage, and nurturing spirit of women.

In this era of globalization, African people face the urgent need to foster world-wide education about their mathematics, science, technology, history and culture (Gerdes, 1998). All groups of people have made significant contributions to the development of mathematics. However, the contributions of Africans other than those related to the early beginnings of algebra and geometry in ancient Egypt, are still highly unacknowledged in the history of mathematics (Lumpkin, 1997). This is partly due to our oral traditions, slavery, colonialism, neo-colonialism, globalization, and the fact that our indigenous mathematics, scientific and technological knowledge as a people is often shrouded in our spirituality. This notwithstanding, efforts must be made to document the mathematical contributions of Africans to enrich the curriculum in a multicultural and interdisciplinary way by providing a wide repertoire of examples of mathematical concepts illustrated from the African context. *God's Own: The Genesis of Mathematical Story-Telling* does this. In particular, it highlights the indigenous mathematical knowledge and cultural beauty of the Ndebele people of Southern Africa.

God's Own: The Genesis of Mathematical Story-Telling is a story about creativity, innovation, and adaptation. It is a story about necessity as the mother of innovation. It is about the philosophy of "do what you can, from where you are, with what you have." It is a story about Nma (Beautiful) Jacob, the survivor, overcomer, successor and faithful servant of God. It is a story that The Absolute Infinite is ever present in our lives shaping our mathematical genomes. Emmanuel – God is with us! His Angels are around us, protecting and guiding us!

Chapter I

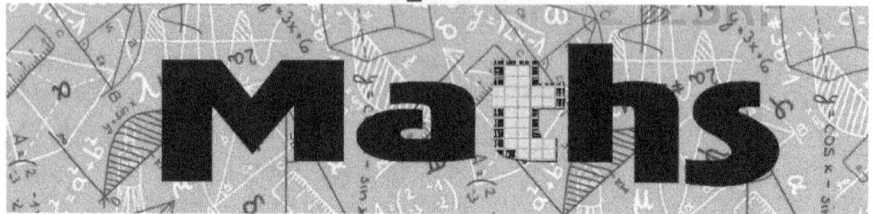

1. Personal Background: I was, I am, My Names, My Childhood Experiences

I am Dr. Nkechi Madonna Adeleine Agwu, aka Nma (Beautiful) Jacob, a widow blessed by God with a successor Ngozichukwuka Jacob Agwu (my teenage son). I was born to Mr. Jacob Ukeje Agwu (father) and Mrs. Europa Lauretta Durosimi Wilson-Agwu (mother). I was born at 10 p.m. on Monday, October 8, 1962, at Parklane General Hospital, Government Reservation Area (GRA) Enugu, Nigeria.

I was almost born on the way to the hospital. My father (Jacob) barely got my mother (Europa) into the hospital when not up to five minutes later I came out "zoom" by normal delivery. I was in a hurry to see the world and begin to chart my physical and spiritual path as Nkechi (God's Own). He is my Jehovah Adonai (Lord and Master). I am to demonstrate His existence in mathematical form as *The Absolute Infinite*. Therefore, He is the cornerstone of my story, and my present scholarship in African culture, and women's stories in science, technology, engineering and mathematics (STEM) related fields. He is the genesis of mathematical story-telling!

I was baptized by my parents (Jacob and Europa) with the names Nkechinyere Madonna Adeline Agwu. My baptism took place on November 4, 1962, at the Holy Ghost Cathedral in Ogui, Enugu, Nigeria. By virtue of the name of this cathedral, Jehovah Nissi (The Lord My Banner) was with me in spirit form. My baptismal registration number is 2936. My baptismal Godparents were Mr. John Iwuanuoge and Mrs. Margaret Asiedu. The officiating priest was Reverend Father Segrave.

My baptismal record and Holy Ghost Cathedral are highly significant because all records of my birth except for my baptismal record at Holy Ghost Cathedral were destroyed during the Nigerian or Biafran Civil War. By the divine intervention of Jehovah Nissi, Holy Ghost Cathedral was not bombed down or destroyed or looted during the mayhem and aftermath of this war. Consequently, I could establish proof of my birth to get replacement records. Unlike then and ancient times, most of the wars or insurgencies in the world today do not respect Holy Ground. Boko Haram, a terrorist jihadist group in my homeland (Nigeria), known worldwide for kidnapping 276 Chibok girls in the night of April 14-15, 2014, from their government secondary school in Borno state, and noted as the world's deadliest militant group responsible for 6,644 deaths in 2014 according to the 2015 Global Terrorism Index, does not hesitate to bomb or carry out mayhem by murdering people in sacred or spiritual places. To date, 219 Chibok girls are still missing. Determine the ratio of the missing girls to the number of girls kidnapped?

War is never pretty. It always brings with it a heavy toll by the shedding of innocent blood, forced human displacement, and destruction of property and valuable records. Leaders of nations, ethnic groups, and clans should always try to avoid making policies and taking actions that could lead their people down that road. They should attempt all peaceable means of resolving differences, including negotiating healthy compromises and agreements. If the *Aburi Accord* of January 4-5, 1967, had been implemented fully as agreed upon in Ghana, the Nigerian or Biafran Civil War could have been avoided.

I was a Biafran child. I was barely five and half years old when the events that led up to the war started. This tragic war caused the death of over two million innocent civilians in Biafra. Many of them were children, women, the aged, and disabled persons. My ancestral village, Alayi, was devastated by Nigerian soldiers during this war. This war rendered my immediate family as displaced persons or refugees for almost four extremely long and painful years of hardship. Prior to the capture of my ancestral village, Alayi, by Nigerian soldiers, my elder brother (Oba) would usually drive us there in the mornings to avoid the day bombings of Umuahia by Nigerian war planes, and bring us back at night. Knowledge of my ancestral village, Alayi, is the only good memory I have of this war. My siblings and I visited Alayi more times during the war than in the entire years of my childhood before the war. It is not surprising that I have strong affinity today for my ancestral village, Alayi.

All our valuables were looted or destroyed during the war or immediately afterwards. In the twinkle of an eyelid, we lost practically all our cherished possessions, birth records and citizenship records. Overnight, I became an "invisible" child. I was "dispossessed" of practically everything that told joyous stories of my birth and my happy life before the war, including access to my ancestral village, Alayi. It was now replaced with tragic and traumatic war stories and memories. In retrospect, the turmoil of the war had a positive influence in shaping my character as a "warrior woman" with an indomitable spirit. This experience of losing significant stories of my life prior to the war is a contributing factor to my passion for story-telling today and my love of history and genealogical mapping.

I thank Jehovah Nissi for His protective covering over my immediate family. We all survived the horror and mayhem of the war without any debilitating physical injury or illness, according to His word in Psalm 23:4 - "Yea, though I walk through the valley of the shadow of death, I will fear no evil: for thou art with me; thy rod and thy staff they comfort me." However, our government assigned home in GRA, Enugu, was completely destroyed. Our personal home in Ibeku, Umuahia, was damaged by bomb fractals while we were living there after forcible relocation from Enugu when this city was captured by Nigerian soldiers.

I remember clearly the day our home in Umuahia suffered severe destruction from a bombing raid. I had to grab my younger brother (Ifeanyi) to run for cover outside our home under the fruit trees we had in our compound. My father (Jacob) loved to eat fruits. He had planted many fruit trees as part of the landscape of our home. These fruit trees, our "Garden of Eden", were Jehovah Nissi's banner over us and even our immediate neighbors. It was where we took cover during the bombing raids in Umuahia by Nigerian fighter planes, which was almost an everyday experience. I believe it was partly the reason why our house was not destroyed like many of the neighboring houses after the war by the Nigerian soldiers. Rather, it was used as an abode by a General of the Nigerian army who was in charge of the management of the people living in Ibeku, Umuahia, and the administration of resources to be supplied to these people, under the 3R's (Reconstruction, Rehabilitation and Reintegration) policy for integrating the former country of Biafra back as part of Nigeria.

This 3R's policy was instituted in 1970 by the Nigerian military government for which His Excellency General Yakubu Gowon was the Head of State. Given the 2015 recent Biafra agitation with non-violent or peaceful demonstrations in the South Eastern region of Nigeria, it appears that after 45 years since the end of the Nigerian or Biafran Civil War, this 3R's policy may not have been effectively executed. One glaring area is the fact that the South Eastern region of Nigeria is the only region with five states despite its high population density compared to other regions that each has six states. This disadvantage in the number of states directly translates to a disadvantage in any federal distribution or allocation of resources made on the basis of states. Hence, the South Eastern region has fewer members in the legislative houses that formulate the laws of the land. The 2014 *National Conference* report approved on March 18, 2015, by the Nigerian Federal Executive Council calls for regional equity in the number of states in Nigeria. This is something that if implemented will demonstrate a true spirit of the reintegration part of the 3R's policy. Hopefully, the current President, His Excellency, Muhammadu Buhari will adhere to the calls by renowned Nigerians for implementation of all the issues addressed within this report.

Both of my parents (Jacob and Europa) were critical war emergency workers for Biafra. Consequently, they were hardly home with us during the day, sometimes even at night, and sometimes for several days on end. Not wanting to come home one day and find their children dead or critically wounded or lost to them because of a bomb raid or some other calamity of war, they chose a viable option of relocating their children to Sierra-Leone with our maternal grandmother, Mrs. Hannah Adeline Wilson (nee Brown). My maternal grandmother (Hannah) was a Sierra-Leonean, a Creole (Krio) woman from Freetown. She was visiting us when the war broke out and she ended up stranded in Biafra. The *Red Cross* was evacuating foreigners stranded in Biafra. My mother (Europa) was a Red Cross volunteer. She was helping to care for and transport out Biafran orphaned children suffering from kwashiokor and stranded foreigners on Red Cross planes to Gabon and Fernando Po (now Equatorial Guinea or Bioko). She was one of those assigned by Red Cross to travel with the refugees due to her ability to speak French fluently, demonstrating the importance of being bi-/multi-lingual. She would travel to villages to pick up abandoned Biafran children sleeping in the forests and bring them to Red Cross refugee camps in Biafra for transportation to Gabon and Fernando Po (Equatorial Guinea or Bioko). She claims we left on the last Red Cross plane out of Biafra to a refugee camp in Fernando Po (Equatorial Guinea or Bioko).

Ironically, I almost did not get transported out of Biafra on this last Red Cross plane with my mother and siblings. When the plane landed at the Uli Airport, to transport those awaiting evacuation, my mother's focus was on helping my younger brother (Ifeanyi) and her aged mother (Hannah) to board the plane. She assumed that I had boarded the plane with my elder siblings when my father (Jacob) told us to board. However, I was so sleepy that I went back into the car when others were not looking to continue sleeping. It appears my spiritual connection to my father (Jacob) was attempting to have me stay behind with my father. I was his Nkechinyerem (His Own Gift From God).

Fortunately, my father decided to check after the family had boarded to see whether we had left anything behind. There he found his Nkechinyerem cuddled up sleeping. He quickly carried me in his arms and ran towards the plane shouting out to the flight attendants "please do not close the door there was one more child to board!" Luckily, they heard his cries and they waited for him to bring me to the doorsteps to board the plane. I sometimes wonder what my life would have been like if I had not boarded the plane, but had been separated from my mother and siblings. Would I have survived the war? How different would my outlook on life be?

My mother (Europa) is my 1st female mentor. Her Red Cross volunteer work caring for and transporting critically ill Biafran children inspires me. It is probably what fostered my initial interest in chaplaincy, missionary, and disaster relief work. I am a trained New York State (NYS) volunteer fire-fighter, trained community emergency response team (CERT) member, and trained NYS chaplain certified to practice globally. I am currently working towards being a fully ordained Christian minister to fulfill my calling as God's Own.

My mother (Europa) is one of the many unrecognized heroes – the silent "women soldiers" of Biafra, who worked hard behind the scenes caring for the orphaned, sick, wounded, the aged, and disabled. The stories of these super women are yet to be fully told so they can get the proper recognition they deserve. Most of the stories of this civil war that have been published are stories told by the male leadership on the Nigerian and Biafran side. A medium term goal of mine is to write the story of my mother (Europa) as a Red Cross volunteer for Biafra. My late husband (Nicholas) worked under her supervision as a Red Cross volunteer for Biafra. He told me a few war stories of her courage and strength that motivated and inspired him.

We were in Fernando Po (Equatorial Guinea or Bioko) for a few months until the British consular office acting on behalf of the Sierra-Leonean government approved our refugee visa entry into Sierra-Leone. My parents (Jacob and Europa) had planned that my mother (Europa) would just see us off to Fernando Po and return to Biafra, since this was part of her regular Red Cross volunteer responsibilities. However, my mother (Europa) was unable to return to Biafra because no more Red Cross planes were flying into Biafra. Also, the British consular office refused to grant us refugee visas to enter Sierra-Leone without our mother (Europa) going with us because they felt that my maternal grandmother (Hannah) was too old to take care of us. We would become a liability to the Sierra-Leonean government. Therefore, not wanting to leave us without an adult guardian in the refugee camp in Fernando Po (Equatorial Guinea or Bioko) to risk returning to Biafra by land, my mother (Europa) made a quick turn-around decision of accompanying us to Sierra-Leone with the hope that she would be able to return later by land to join her husband (Jacob) in Biafra, after we were settled. Through this experience, I learned from my mother (Europa) that to love your child means to sacrifice for the well-being of your child, even if it means placing your child's needs above your spouse's needs.

This dogmatic policy of the British consular office that forced a change of plans for my mother (Europa) turned out to be the blessing of Chukwu (God) in disguise. We encountered tragedy in Sierra-Leone that really required my mother's presence. This prevented her return to Biafra until several months after the war had ended when she went back to locate her husband (Jacob). Through the grace of Chukwu, she was able to fund her travel from Sierra-Leone to Nigeria through her consulting work as a West African Examinations Council (WAEC) examiner, one of many consulting jobs she took on for additional income to care for her family. She was also a consultant for the United States Peace Corps helping to provide professional development for those who went to teach in schools in various locations in Sierra-Leone.

WAEC paid for my mother (Europa) to attend a meeting in Lagos, Nigeria, in the later part of 1970, the year when the war ended. After this meeting, she used the opportunity to travel to Ala Igbo (Igbo land) to search for her husband (Jacob). Thanks to Chukwu (God), she found him in our home at Ibeku, Umuahia, alive and well. He was struggling to survive and take care of his mother, Mrs. Omamma Virginia Agwu (nee Egu Agu) and other members of our family clan of Umuapu, Agbakoli, Akoliufu, Alayi. This hardship caused us to stay much longer in Sierra-Leone after the war had ended to allow my father (Jacob) sufficient time to get back on his feet financially.

Living in the refugee camp in Fernando Po (Equatorial Guinea or Bioko) taught me several valuable lessons at an early age about survival and self-preservation. Sadly, I also learned what it meant to be a "street" child. I have never been back to Fernando Po. I hope to take my son (Ngozichukwuka) to Fernando Po (Equatorial Guinea or Bioko) before I leave this earth. My stay in the refugee camp there fundamentally shaped my character of perseverance, determination, and imbibed in me the spirit of humility. We left Fernando Po (Equatorial Guinea or Bioko) by ship to another refugee camp in Monrovia, Liberia. We were bundled in the ship like "sardines" in the worst quarters with the worst possible meals with restricted access primarily to our quarters because we did not pay to be transported. This is how I developed *sea sickness* and a fear of water which I eventually overcame much later in my adult years. Through this experience, I can envision some of the sea experiences of slaves of ancient times that were transported like cargo to Europe and the Americas during the *Atlantic Slave Trade*.

Finally, we left Liberia via a plane chartered by the Sierra-Leone government to another refugee camp in Freetown, Sierra-Leone, until we were re-settled. We arrived in Sierra-Leone on October 4, 1968, just a few days before my 6th birthday. This was the only year when growing up as a child that my birthday was not celebrated with a party. That was due to our circumstances of homelessness. However, I thanked Chukwu (God) and praised Him, according to His word in Psalm 34. On this particular birthday, I celebrated life and good health - escaping from the mayhem, turmoil, and trauma of war. This 10th month of the year (October) is a highly significant to me. It is the month of my birth. It is the month of my son's birth. It is the month I arrived Sierra-Leone as a Biafran War refugee. It is the month I was widowed. It is a month that either reflects great joy or sadness – the *Agwu* Igbo philosophy of opposing forces, depending on the lens you use to look through.

Agwu (my family name) refers to the Igbo patron deity of healing, divination, inspiration, and talent, charged with the responsibility of shaping one's destiny. It also refers to a basic Igbo theological concept used to explain good and evil, health and sickness, wealth and poverty, fortune and misfortune – opposing forces to everything.

Ever since I learned the meaning of Agwu, I have been interested in its origin as our family name and the reason why my father's ancestral line in the only lineage in Ndi Mbada that bears Agwu as a family name. When my father (Jacob) was alive, I used to question him about this issue. He would artfully dodge answering my questions. I have conjectures about why my father (Jacob) who was passionate about Alayi history and family genealogy would become the artful dodger on the topic of the origin of our family name. I believe it has to do with traditional Igbo beliefs that there are Agwu people in every community.

Agwu people are recipients of the deity's positive influences because the deity has endowed them with gifts as healers and diviners or they are victims of the deity's malignant powers because of their refusal to accept its call. According to Igbo traditional beliefs, Agwu (the deity) choses who it wishes to endow with its gifts in a family whose ancestral line had someone previously endowed with its gifts. I am still investigating the origin of our family name and Igbo traditional and cultural beliefs about Agwu.

Re-settling in Sierra-Leone took a few months as we were faced with homelessness once again. My maternal grandmother's home at Pademba Road in Freetown that my parents (Jacob and Europa) had hoped would be available for us to reside in and provide for us through rental income had been accidentally burned down while my grandmother was in Nigeria. The letter that was sent to her informing her of this sad news never arrived in Biafra due to the war. No one really insured their homes in those days, so there was no insurance money for my maternal grandmother (Hannah) to use in re-building her home. Neither did she have any significant personal funds of her own to do so. She could not seek a bank loan to enable her to do so either as she was now unemployed.

Unfortunately, none of our immediate relatives, my mother's sisters, could assist us with a place to stay as they themselves were struggling and living in cramped quarters. My mother came from a low-income background. My maternal grandmother (Hannah) was a petty trader and market woman. She was married to a Sierra-Leonean Mr. Jacob Wilson, a Creole (Krio) from Freetown. He was a former employee of the Nigerian Government in Jos, Nigeria. Jos is where my mother (Europa) was raised after she was born on April 30, 1927 in Freetown, Sierra-Leone. My maternal grandmother (Hannah) had gone back to Freetown, Sierra-Leone, to deliver her. My maternal grandparents (Jacob and Hannah) did not consider the hospitals in Jos and its neighboring environs to be good hospitals for providing adequate care for delivery of babies.

My maternal grandfather (Jacob) died when my mother (Europa) was barely six years old, leaving my maternal grandmother (Hannah) a widow early in life and causing her to relocate back home to Freetown, Sierra-Leone. The only two sons of my maternal grandmother (Hannah) also died leaving their wives as widows early in marriage. All her daughters except my mother (Europa) were widowed early. Being widowed early (under age 50) or married but living separately from one's spouse for many years, such as in my case, appears to be an unseemly pattern in my generational line both on my paternal and maternal side. By my "faith as small as a mustard seed", according to Psalm 91:16, my son (Ngozichukwuka) and all who will come through his lineage shall never inherit this life pattern, in the mighty name of Jesus. Amen.

From the refugee camp in Freetown, being homeless, our family had to be separated to live with family friends and relatives of my mother (Europa) who were not her siblings. My mother (Europa) and her two youngest children, my younger brother (Ifeanyi) and I, stayed with the Leigh family at Congo Cross. Others stayed elsewhere. My parents (Jacob and Europa) became friends with Mr. and Mrs. Leigh during the period prior to the civil war when Mr. Leigh was working for the Nigerian Railroad Corporation. They were extremely kind and generous towards us during the period they sheltered us. My mother (Europa) is the only member of her immediate family to attend university. She is a 1st generation university graduate. She has two degrees at the Masters level in Mathematics Education and School Administration. So, she was able to get a good job in a vacant position as a senior teacher and boarding school mistress at the Freetown Secondary School for Girls (FSSG), Freetown, Sierra-Leone, a few months after our arrival in Freetown. This helped her to take care of her family and to provide us with a roof over our heads. She later became a Vice Principal of FSSG. I learned first-hand from this experience the value of having a good education. I was determined that I would exceed the educational level of my parents by pursuing a doctoral degree in whatever field I selected for my career.

I originally wanted to be a medical doctor and science researcher, in particular, a neurosurgeon and researcher in the field of neuroscience. I am interested in the functions of the brain, how it works, and how it facilitates knowledge construction and learning. Interestingly, my son (Ngozichukwuka) would like to become a neurosurgeon and if not, a veterinary surgeon. He decided on neuroscience as a career interest without any knowledge of my own career interest as a child or any urging on my part. I changed my mind about becoming a medical doctor in form four in high school after witnessing live in the operating theatre three surgeries by one of the most renowned surgeons in Sierra-Leone, Dr. Ulrick Jones. One of these operations was an extremely bloody brain surgery for relieving cranial pressure suffered by the patient. I had been shadowing Dr. Jones during my vacation period from high school while I was awaiting breast surgery for Fibroid Adenoma to be done by him. I just could not stand the sight of all the blood I saw in the operating theatre that day. It brought back vivid memories of wounded and dying Biafrans during the war.

I once again experienced these memories when running for survival upon evacuating Fiterman Hall. This is a building owned by my university, Borough of Manhattan Community College, City University of New York (BMCC CUNY). It is where I was teaching September 11, 2001, the day of the terrorist attacks of the *World Trade Center* (WTC). I had to question myself seriously on whether I would be able to overcome this aversion for the sight of blood to excel as a medical doctor. I decided I was not ready to put it to the test. I settled for my next best love. This was mathematics. It was a subject I excelled in throughout my school days.

I am a teacher today because of my love for the profession, perhaps from years of watching my parents as teachers. If I had let my parents (Jacob and Europa) select my career, I would have been a medical doctor, engineer, architect, lawyer, or a professional in one of those highly paid professions. Although my parents had both served in the honorable teaching profession and loved it, they did not want it for their children because it was not as financially lucrative as the other professions mentioned above. I never became a neurosurgeon or a researcher in neuroscience. However, I am practicing a form of "pseudo" cultural neuroscience with my scholarship in African culture and women's stories in STEM.

Ironically, my home countries, Sierra-Leone, United States, and Nigeria, have each had their own share of war or acts of terrorism that have tragically affected me. Sierra-Leone underwent a long 10 year civil war in which we once again lost cherished possessions and important documents that told stories about our life in Sierra-Leone. Rebel soldiers burned to the ground the home of my aunt in Circular Road, Freetown, where these possessions were being stored since my mother's home in Congo Cross, Freetown, was rented out. This war caused the death of my mother's sister and favorite aunty Elizabeth, popularly known as Eliza, when her home was burned down by the rebel soldiers. Also, my university (BMCC CUNY) lost the use of Fiterman Hall during the September 11, 2001, WTC terrorist attacks. One of the twin towers of the WTC, building seven, partially collapsed on Fiterman Hall. It took several years before my university was able to re-build Fiterman Hall. For a few years after Fiterman Hall was rebuilt, I avoided teaching or going there because it brought back tragic memories of that day. Additionally, we lost numerous members of our extended family from Ndi Mbada, Agbakoli, Akoliufu Alayi, Nigeria, through the civil war in Nigeria. My favorite uncle and mentor (Robert) who lived with us in Enugu when the war broke out died as a child soldier fighting for Biafra. I can remember how proud my uncle Robert was to join the Biafran army. These war and terrorism experiences have taught me to understand clearly how things and people that you cherish can be ripped away from you in the twinkle of an eye – here today, gone tomorrow forever. It is extremely important that you let your loved ones know every day how important they are to you. You never know whether that may be the last time you will see them alive. These experiences have also brought me closer to God, have kept me in tune with my spiritual side, and have helped me to be more humane and Christ-like.

Although my father (Jacob) was not the most senior surviving male figure of our family clan of Umuapu, by default of being the most educated, globally exposed, and professionally accomplished man in our clan, he held the mantle of leadership. Overnight, he became a husband to many of the widows, not in the sense of physical intimacy, but by caring for them, helping to raise their children, supporting their work endeavors as market women, farmers and so on, and building homes in Ndi Mbada for those whose husbands died without building them a home. I am neither male nor the most senior of his surviving children, but it appears this torch has passed on to me. I am carrying it with a spirit like Dorcas and sacrifices like the widow of Zarephath in the Bible. I am uniquely positioned to maintain this tradition of my father (Jacob) because I understand their plight, being a widow with a successor.

In traditional Igbo life, a person's name provides insight into their life and reality. This is why my story begins with the stories of my names. The privilege of which name the community will use to address a child is given to the parents or grandparents. This provides an avenue for them to make a significant statement related to their life experiences and to express deeply felt wishes on their hopes and expectations for the child. The naming ceremony usually takes place on the fourth or eighth day depending on the health of the mother and child. The numbers four and eight are significant numbers in the Igbo culture. They are related to the Igbo calendar and numeration system. I plan to conduct further in-depth research on the significance of these numbers. The four-eight day cycle serves to synchronize the inter-village market days. Interestingly, I have a four-eight day number pattern in my life, my baptismal date – November 4, 1962, and my birth date – October 8, 1962.

My paternal grandmother (Omamma) named me Nkechinyere (Gift Of God). My father (Jacob) always called me Nkechinyerem (My Own Gift From God). He was implicitly and repetitiously telling me that my life signifies a spiritual connection between him and God, and that I have an important assignment of his to fulfill. I now know my assignment.

My father (Jacob) was the only surviving and last born child of the six children of his father, Mr. Ukeje Udegbe Agwu and his mother, Mrs. Omamma Virginia Agwu (nee Egu Agu). His five siblings died of unidentifiable causes at infancy within the first six months of birth. He suspected it was due to *Sickle Cell Anemia* given that both him, and his mother (Omamma), were carriers of the Sickle Cell genes. His dead siblings were termed by the Alayi community "nwangbala uwa", the child who comes with no intention of staying.

The infant mortality rate in Nigeria then was extremely high. It still is. It is 158/1000 according to the 2011 data of the United Nations International Children's Education Fund (UNICEF). Write this ratio in reduced form? Research the UNICEF infant mortality rate for other countries in Africa and around the world, and write a comparative analysis. There is plenty of work in public health education to be done by the medical community in Nigeria to minimize these statistics on infant mortality and the death rate for children as a whole. The Ministry of Education in Nigeria may want to develop curriculum that will promote genealogical mapping and public health education. I have a template in the area of public health which the ministry could adapt given my current work developing curriculum for teaching introductory statistics, with a colleague in the department of science at BMCC CUNY, Professor Christopher Salami, who was a former Lecturer at the University of Benin in Edo state, Nigeria.

The maternal grandfather (Egu Agu) of my father (Jacob) died 18 months prior to his birth on August 16, 1925. So, my father (Jacob) was considered by the Alayi community as the re-incarnation of Egu Agu or the child who came to console his mother on the loss of her father. His father (Ukeje) passed away when he was only four years old. His mother (Omamma) made a wise decision to send him to be raised by an Irish missionary, Reverend Father David Walsh, while serving God as an altar boy in the Catholic Church. Reverend Father Walsh sponsored his primary school education at Saint Joseph's Catholic School, Alayi. Due to his excellent performance there, Father Walsh encouraged him to consider a career in teaching. He became a probationary teacher in his alumni primary school, teaching mathematics and geography, until he left for further education at Saint Charles College in Onitsha, Nigeria, to earn his Nigerian Higher Elementary Teacher's Certificate and London Matriculation.

As a child, I was a worshiper of God in the Catholic tradition because of how my father (Jacob) was raised. Now I no longer do so. At my current institution, I was blessed to meet a highly significant spiritual and professional mentor, Bishop (Dr.) Joe Omeokwe. He is a retired faculty member and former Mathematics Department Chair at Touro College, New York, United States (US) and the Archdeacon of the Anglican Missionary Diocese of the Trinity, North East region. Through his spiritual mentoring since the birth of my son (Ngozichukwuka), I now worship God in an interdenominational way at His Vineyard International Christian Ministries (aka The Harvest Center) in the Bronx, US, a church founded by Bishop Omeokwe. I also worship God at His Glorious Miracle Embassy International (aka Power House) in Umuahia, Nigeria, under the spiritual covering of the founder, Prophet (Rev.) Emmanuel Angel.

I was confirmed as a child at Saint Anthony's Catholic Church in Brookfields in Freetown, Sierra-Leone. I chose Mary Magdalene as my Catholic confirmation name because she loved Jesus and was devoted to Him. I am inspired by the consistency in her devotion to Jesus even after his crucifixion. I aspire to emulate this type of love for God as His own (Nkechi).

I am a noble woman both by name and the mantle of family clan leadership that rested on my father (Jacob). Adeline (Noble) is a traditional family name on my maternal side. It was the name of my maternal grandmother (Hannah). She is the angel that God used to relocate us to safety during the latter part of the war. I found out when we relocated to Sierra-Leone that I share the name Adeline with many of my cousins. In order to be unique or differentiate myself from all my cousins who bear this name, I changed the spelling of my name to Adeleine, which includes an additional "e" after the "l" and gave birth to a new variant of the name. My boldness in breaking away slightly from the spelling of this traditional family name tells you something about my character. I am not afraid to stand my ground or stand alone, even where there may be grave consequences for doing so. I have undergone a few negative fallouts for my courageous walk in this regard. However, Jehovah Nissi has always come through as a banner to preserve me.

My nobility is also evident in my name Madonna (My Lady). By this name of mine and the color of my skin, I am a "Black Madonna", an appellation of the Virgin Mary. I am clothed from head to toe like an *Ndebele* doll wrapped with a beautifully embroidered garment of Africa, woven by rural Nigerian women and the Nigerian Women in Agricultural Research for Development (NiWARD). I have been painted and sculptured in alternative forms of representation. The form I love most is that which births the decade of recognition, justice and equity for the Global African Woman. This form celebrates the work of the late Dr. (Mrs.) Mojisola Edema, a founding mother of NiWARD and the indigenous mathematical knowledge of rural women in Nigeria who weave traditional clothing such as Akwete cloth. Produced by the Drammeh Institute is a YouTube video of this representation at:

https://www.youtube.com/watch?v=dkycEMQ0fyg.

Like my parents (Jacob and Europa) I am a mathematics teacher, educational pioneer, and a mentor to many children, youth and adults. We share parallel lives or symmetric triangles in this regard. My father (Jacob) is the 1st university graduate of Alayi. He is a pioneer for his research and professional endeavors in agriculture and rural development in Eastern Nigeria. Two highly significant publications of his are *Ruralism: The Correct Approach to Economic Independence of Abia State* and *An Experiment in Rural Development in Eastern Nigeria: The Case of the Akoliufu Pilot Project.* My mother (Europa) is the 1st female university graduate of mathematics in Sierra-Leone. She is one of the Sierra-Leonean pioneers for mathematics textbook writing, with the Longman's textbook, *Congruence and Constructions: A Unit for Secondary School Mathematics,* Teachers Edition, Volume I. She authored this book in Arithmetic and Geometry with fellow colleagues. I am the 1st female doctoral graduate of Alayi. I am a pioneer in mathematical biographical story-telling using vertex-edge graphs and number patterns.

My parents (Jacob and Europa) are benevolent people. They raised and educated a "nation" of children and youth, a few whom they officially adopted, educating me first-hand that one of the best investments a person can make is investment in the education of others. They were addressed by all whom they raised simply as "Daddy" and "Mummy". I am sometimes conflicted in my count of how many siblings I have. Therefore, rather than identify a specific number of siblings that could be debatable depending on your lens, I will identify myself as the youngest daughter among all the children they either gave birth to or officially adopted.

Ever since God called away my father (Jacob) on June 7, 2008, I find myself walking more and more in his footsteps even though his shoes are too big for me to fill. It appears I have taken on his persona or that his "invisible" hands are directing my footsteps from where he is laid to rest in the estate of *Jacob's House*, Ndi Mbada, Agbakoli, Akoliufu, Alayi, Nigeria.

Today, I answer the shortened form of my name Nkechi instead of Nkechinyere because the meaning of Nkechi (God's Own) reflects how I now see myself. Many people foreign to the Igbo culture find it difficult to pronounce Nkechinyere. Also many of my close friends prefer to call me by the shortened form or even shorter forms of Nke (Gift) or Nk or some nick name such as "Nky" or "Nky babe". Moreover, my paternal grandmother (Omamma) and father (Jacob) who are the two people in my immediate family that addressed me by the long form of my first name have been called away by God. Other members of my immediate family, including my mother (Europa) have always preferred to call me Nkechi.

I use the alias Nma (Beautiful) Jacob for my creative work (mostly poetry) because of what it signifies. Nma is derived from the first letter of each of my names. It is an Igbo name that means beautiful or good and it is the shortened form of my paternal grandmother's name (Omamma). I was created by God beautiful because I am formed in His like image, Genesis 1:27. Jacob (Successor or Overcomer) is a traditional family name both on my maternal and paternal side. It is the name of my father, my maternal grandfather, my son, and name of many male members of my extended family both on my maternal and paternal side. I am a successor of my father. I am an overcomer. Therefore, I am Jacob. Also, in Ala Igbo (Igbo land) during ancient times, people answered their father's first name as their last name. By answering Jacob after Nma, I follow the tradition of my culture.

This alias Nma (Beautiful) Jacob metaphorically represents my scholarship in the area of African culture and women's stories in science, technology, engineering and mathematics (STEM) related fields. Jacob is cultural (family and Igbo tradition in naming) and Nma is about women's stories (hidden stories about beauty, nobility, character, spirituality, charity, love, goodness, and much more). By using Nma (Beautiful) Jacob as an alias for my creative work, I memorialize my father (Jacob) who taught me valuable lessons about creativity. I give honor and sing praises to my Father Jehovah Elohim (God The Creator), the God of Abraham, Isaac, and Jacob, the God of Elizabeth, Sarah and Hannah. He created each of us as beautiful beings with the ability to be creative when He formed us in His like image. I have found my own creative Nkechinyerem (My Own Gift From God).

Chapter II

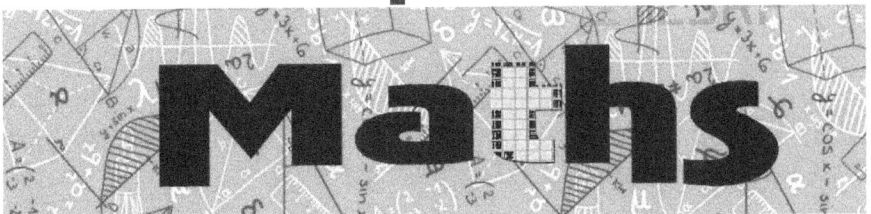

2. My Persona: A NiWARD Talking Drum, Shekere Musical Instrument, and Akwete Cloth

It is not by chance or accident but by the divine intervention of The Absolute Infinite (God) that I find myself situated as a Nigerian Woman in Agricultural Research for Development (NiWARD) story-teller. I am the daughter of a 9th generation farmer from Umuapu, Agbakoli, Akoliufu, Alayi. My father (Jacob) educated me formerly and informally through his engagement in agricultural research and development, within the context of rural development in Nigeria. To understand my identity and my philosophical approach towards NiWARD you need to know about my father (Jacob). He is the cornerstone for the Temple of Nma (Beautiful) Jacob or the Temple of Nkechi (God's Own).

My father (Jacob) was thrice a First Bank *Farmer of the Year* honoree. His agricultural and engineering company JANES Nigeria Limited Incorporation was awarded a sponsorship by the Standard and Chartered Bank to exhibit his agricultural engineering innovations at the *Royal Agricultural Show* in Britain in 1973. He was a founding father of Alayi Farmers' Cooperative and a founding father and first Secretary General of the Alayi Development Union (Ndi Oko Anyi), founded in 1948. He was a philanthropist. I grew up to understand what it means to be benevolent by watching him in action. In 1964, he established a 510 hectare oil palm plantation which he donated for the prosperity of Alayi through the formation of Amaeke Na Agbakoli Multi-Purpose Cooperative Society Limited, an organization with the mission of poverty alleviation and employment creation in Alayi.

Prior to the Nigerian or Biafran Civil War, my father (Jacob) was a civil servant. He served Nigeria as a Permanent Secretary for Rural Development for the Eastern Region and before that, as a diplomat at the Nigerian Mission in the United Kingdom. Prior to that, he was an employee of Shell Oil. My father (Jacob) quit his Shell Oil job to serve his country at less than half his Shell Oil salary when he discovered inequitable or discriminatory salary practices at Shell Oil. According to him, Shell Oil employees hired outside Nigeria, even those who were Nigerians hired in other African countries such as himself, were earning higher salaries than their counterparts at the same level. This change in employment for a lesser salary was one of his many life examples of sacrifice for his country and standing up for the principles you believe in.

My father (Jacob) served on the Biafran national policy-making cabinet for science. He was a core member of the scientific team that developed essential machinery and equipment used during the war, such as the deadly missile known as the *Ogbunigwe*. These scientific and technological innovations by the Biafran scientific team served to prolong the war. It is unfortunate that the Nigerian government has not harvested the creative knowledge that led to these innovations in their efforts to promote science and technology in Nigeria, and to foster creativity and innovation in the education of Nigerian children. Rather, the preference of the Nigerian government is for this knowledge to die with the few remaining Biafran scientists who are still alive as part of their efforts in suppressing the history of Nigeria pertaining to the civil war. It is for this same reason that the history of the civil war is excluded from the primary and secondary school educational curriculum, even though it can be used to teach many lessons about conflict resolution, peace negotiations, bridge-building and the Geneva Conventions.

My father's leadership role in Biafra during the war resulted in him receiving the *Biafran Silver Medal*. However, it cost him his employment in the Nigerian Civil Service after the war through compulsory retirement by the Nigerian government. It forced him to go into hiding at the end of the war to avoid being killed by Nigerian soldiers. Consequently, he became an "accidental" business entrepreneur, doing what he knew best, farming, as a way to provide for his family (immediate and extended) and all who depended upon him for survival and an education.

My father (Jacob) was Chief Executive Officer (CEO) and Chairman of the Board of Directors of JANES Nigeria Limited Incorporated, an agricultural and engineering company he founded after the war as a means of survival for himself and his entire "village" or "nation" of dependents. It is through his salary from JANES Nigeria Limited Incorporated as a farmer that my siblings, many of those he educated, and I, were able to begin our higher education journey. I was hoping to come back home to Nigeria to join him at JANES Nigeria Limited Incorporation and to support his philanthropic work in Alayi, once my son (Ngozichukwuka) completes his undergraduate education. However, God called him away before that could happen. God knows best the reason why. In all things I give thanks to God.

During my holidays from university, as an undergraduate student majoring in mathematics at the University of Nigeria, Nsukka (UNN), I occasionally helped out on the poultry farm of JANES Nigeria Limited Incorporated at Ikot Ekpene Road in Ibeku, Umuahia, Abia State, Nigeria. I was learning the ropes of rearing chicken for mass production, and administering and managing a poultry farm. I got tired of eating eggs for breakfast and chicken with my lunch and dinner. It was easily accessible "easy fare" to cook by household members. We did not have to go to the market to buy meat. I have adopted this easy fare with my approach to teaching students to create, play, and engage in real-life mathematics with the Okwe (Mancala) game created out of an empty dozen egg carton and 48 seeds. It reminds me of the poultry farm of JANES Nigeria Limited Incorporated. I sometimes use beans for the seeds of the game. The beans are a reminder of my daily diet in the Red Cross refugee camp in Fernando Po (Equatorial Guinea or Bioko). For many years, I disliked eating beans because I ate too much of it at this refugee camp.

I occasionally shadowed my father, Jacob, in his JANES Nigeria Limited Bende Road agricultural engineering research laboratory in Umuahia, Abia State, Nigeria. Unfortunately, this laboratory where he developed his three patented inventions, *The Rapid Oil Mill*, *The Hand Operated Horizontal Oil Press Mill*, and *The Rice Husk Ash Cement Block Press*, was taken over without compensation for the land and for the relocation of the laboratory and its equipment, by the Abia State Government to serve as the state police headquarters when the state was newly created.

God's Own: The Genesis of Mathematical Story-Telling

This type of land seizure demonstrates the value that the then policy-makers placed on fostering an enabling environment for scientific innovators or for agricultural engineering, which was the focus of his laboratory endeavors. Hopefully, federal and state legislators in Nigeria will enact legislation that prevent the federal or state governments from taking actions that close down legitimate business laboratories that are well documented as facilitating scientific innovation for agriculture and rural development.

At this Bende Road laboratory, my father (Jacob) was constantly tweaking agricultural products of palm oil, garri, rice, and cashew nuts to formulate and test his conjectures and hypotheses, and devising adaptions of machinery to process them better and more efficiently. In so doing, he taught me the basics of scientific investigation and demonstrated to me first-hand what creativity, innovation, and adaptation was all about – "do what you can, with what you have, from where you are." This has become my research mantra more especially because of the difficulty in obtaining funding or support to move forward my virgin, innovative, creative and interdisciplinary scholarship in African culture and women's stories in science, technology, engineering and mathematics (STEM) related fields.

This ethno-mathematics scholarship bridges the sciences with the humanities and the arts. Most funding sources, even those that fund interdisciplinary projects, tend to be compartmentalized either to the sciences, the humanities or the arts. They tend to overlook virgin or new areas that cut across the sciences, the arts and the humanities. So pioneers in this regard never have it easy or go unscarred. I am finding this out first-hand. However, my eyes continue to stay focused on the mission of this work and not on the disappointments from inadequate funding or other professional set-backs. I have faith that God will "open the Heavens" on my behalf to preach, teach and demonstrate the gospel of His existence as The Absolute Infinite with this scholarship of mine in African culture and women's stories in science, technology, engineering, and mathematics (STEM) related fields.

Daddy, I thank God for giving you wisdom to imbibe in me your spirit of creativity, innovation, and adaptation. I am trusting in God that I will pass it on to your successor, Ngozichukwuka (my son). You identified him at an early age of 18 months as having a creative mind when you watched him tinkering with your stuff in our family home in Ibeku, Umuahia, trying to create something out of nothing. You sent for me to come and watch what he was doing. You advised me then to always nurture his creative spirit. I am trying to do this.

Daddy, I thank God for this opportunity to tell a little bit of your genius story through its reflection in my life. A medium term goal of mine as a biographer is to complete your autobiography so that our African children can see your name among the list of African geniuses. You started writing it with my encouragement but could not finish as God called you away from all those who loved you. If you were from in a developed nation, your name would probably be ranked next to the Einstein's and Edison's of this world. Your genius would have been recognized early and provided with the type of nurturing that a rose needs to fully blossom. You would have had more than your three patents for agricultural engineering innovations. You operated under the philosophy that agricultural waste can always be recycled for use in other beneficial ways. I have equally imbibed this philosophy that you can always create something out of waste. It is the reason why the Ndebele doll activity which I use to teach concepts in graph theory and number theory using stories of women from the group of NiWARD in my writing intensive discrete mathematics classes at BMCC CUNY is one of my favorite cultural curricular activities. The Ndebele dolls of NiWARD are created out of recycled materials.

From the mapping of my family genealogy tree vertex-edge graph, I am a 10th generation farmer from Umuapu, Alayi, Nigeria. However, unlike my ancestors, I plant my seeds in humans. By God's divine intervention, I have found a specific path that I must follow in planting seeds in humans to bring recognition to the traditional work of rural women in Nigeria, to bring visibility to the successful women of NiWARD, and to build capacity in science, technology, engineering and mathematics (STEM) related fields. This path requires the development of curricular activities framed around African culture and women's stories. Therefore, I am a NiWARD *Talking Drum*, *Shekere* musical instrument, and *Akwete* Cloth!

God's Own: The Genesis of Mathematical Story-Telling

My assignment is to use the Talking Drum to send out curricular coded messages about NiWARD, the work of rural women in Nigeria and the indigenous mathematical knowledge inherent in Nigerian cultures. The Talking Drum is already in employment sending out test messages to the world at large. One of these messages that relates to the work that was done in August 2014 on our Carnegie African Diaspora Project (CADF), *Culture and Women's Stories: A Framework for Capacity Building in Science, Technology, Engineering and Mathematics (STEM) Related Fields,* is published at:

http://www.hostos.cuny.edu/MTRJ/archives/volume7/issue2/volume7issue2full.pdf.

Another message celebrating the work of the late Dr. (Mrs.) Mojisola Edema, a founding mother of NiWARD, and highlighting the indigenous knowledge of rural women who weave traditional cloth such as Akwete and Aso-oke is published at:

https://www.youtube.com/watch?v=dkycEMQ0fyg.

My assignment is to use the Shekere to play the agricultural research and development music of the numerological and other number patterns of NiWARD, just like I have done for myself in the later part of my story. My assignment is to use the Akwete Cloth as a tapestry to illustrate the beauty and uniqueness of the mathematical genome and vertex-edge graphs stories of NiWARD. Pablo Picasso did likewise in illustrating the mathematical beauty of African cultures with his African influenced art in the Proto-Cubist period of 1906-1909. This is my Nkechinyere (Gift Of God).

As a typical farmer, I trust Chukwu (God) for good weather and all other conditions that will yield a great harvest in terms of capacity building in STEM. He said in Matthew 7:7, "ask and it will be given to you." I have asked this of Him at the altar in His Vineyard International Ministries (The Harvest Center) in the Bronx, United States. I shall celebrate the next Igbo New Yam Festival giving thanks to Him and showing off the fat fertility yams I have harvested in the lives touched by this project as a NiWARD Talking Drum, Shekere musical instrument and Akwete Cloth at Jacob's House, Ndi Mbada, Agbakoli, Akoliufu, Alayi. I shall give thanks and glorify Him for this great work of His at His altar at Glorious Miracle Embassy International (Power House) in Umuahia, Nigeria.

The New Yam Festival is significant to me because my father (Jacob) was born on the Sunday, August 16, 1925, of the *Ikeji* week, the week of the New Yam Festival. When I celebrate this festival, I am celebrating his life. Incidentally, August 16, 2008, is the day I birthed the name Nma (Beautiful) Jacob to celebrate the life of my father (Jacob) in the year of his death. I am working earnestly towards this short-term goal of showing off my fat fertility yams at the next New Yam Festival in 2016. I am doing so with three sets of students viz:

- 12 disciples in the discrete mathematics class I am currently teaching at the Borough of Manhattan Community College (BMCC), City University of New York (CUNY) for the 2015 Spring Semester.
- Two female students whom I am mentoring in the Exploratory Science Research class at Pelham Lab High School in the Bronx as part of their Science and Technology Entry Program (STEP) research requirements of their school's partnership program with the State University of New York (SUNY), Albany.
- High school students in the BMCC CUNY Fall 2015 STEP Program.

This classification of students, are serving temporarily as my agricultural nursery seeds. My agricultural nursery is presently at BMCC CUNY. It shall be relocated at a later date to Nigeria to Jacob's House in Alayi. This is where it appropriately belongs. Jacob's House in Alayi is a place of Rehoboth. It shall be a place for nurturing and grooming young disciples (children and youth) of Nma (Beautiful) Jacob, in African culture and women's stories in STEM, and teaching them about the existence of The Absolute Infinite (God), according to Genesis 26:22, in the mighty name of Jesus. Amen.

God's Own: The Genesis of Mathematical Story-Telling

My favorite childhood fairy tale was *Cinderella* because she overcame severe challenges before meeting her prince charming with the help of her fairy Godmother. I am now like "Cinderella". I have overcome several professional challenges to meet my prince charming, NiWARD. These challenges also include challenges related to applying for and implementing our Carnegie Project, as well as, continuing it beyond the Carnegie Foundation funded period. However, Jehovah Nissi (The Lord My Banner) has always provided His protective covering to overcome these challenges.

Professor Emeritus Stella Williams is my Fairy Godmother. She discovered my hidden African traditional story-telling talent as a Talking Drum, Shekere musical instrument and Akwete Cloth at the 58th session of the United Nations Commission on the Status of Women parallel forum on the *African Woman and STEM*. You can find out about my practice in African traditional story-telling at:

https://www.youtube.com/watch?v=CdWar62RrwE.

I thank God for this His angel, Professor Emeritus Stella Williams, who He used to bring me to the NiWARD Ballroom. She is nurturing and mentoring my scholarship related to NiWARD. I thank God for all the NiWARD Disciples that are making this curricular project possible through their stories. In particular, I thank God for my Federal University of Technology, Akure (FUTA) Carnegie Project collaborators Ms. Olabukunola Williams - NiWARD Coordinator, Dr. (Mrs.) Olayinka Ogunsuyi – Past Acting Director of the Center for Gender Issues in Science and Technology (CEGIST), and the late Dr. (Mrs.) Mojisola Edema - a founding mother of NiWARD and Past Acting Director for CEGIST, who worked tirelessly with me on this project. God shall continue to strengthen his NiWARD Disciplines in this initiative.

Specifically, the late Dr. (Mrs.) Mojisola Edema is my NiWARD Prophet. She helped me root myself firmly as a vertex-edge graph African story-teller. Her presentation on our Carnegie Project at FUTA on August 26, 2014, likened my vertex-edge graph stories to the bed-time stories of Maryam Mirzakhani which she used to travel the world in her mind when she was eight years old and to her thinking that mathematics research is like writing a novel. It was as if a light bulb had clicked somewhere. This was my *Eureka* moment!

Maryam Mirzakhani, is a Professor of Iranian heritage at Sanford University, California. She is the 2014 *Fields* Prize Medalist of the International Congress of Mathematicians. She is the 1st female to win this medal. As seen through the eyes of the late Dr. (Mrs.) Mojisola Edema, this NiWARD mathematical work will join the ranks of works considered for some renowned international honor or fellowship of similar caliber to the Fields Medal or MacArthur Fellowship. Her labor of love toiling the NiWARD farm for food production together with our sister, Professor Emeritus Stella Williams, also a NiWARD mother shall be rewarded and shall be celebrated on an international level.

The 8th month (August) like the 10th month (October) is a highly significant to me. It is the birth month of my father (Jacob). It is the birth month of my alias (Nma Jacob). I defended my innovative American Educational Research Association, Division K, and Syracuse University (SU) Creative Research Award winning doctoral thesis titled, *Using a Computer Laboratory Setting (CLS) to Teach College*, at Syracuse University (SU), Syracuse, New York, on August 4, 1995. It is the month in which the Igbo people celebrate the New Yam Festival. It is the month of my Eureka moment!

Chapter III

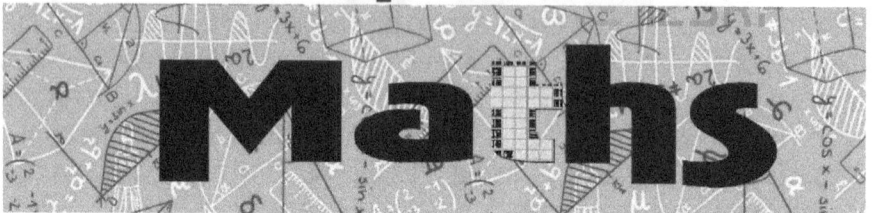

3. Journey to be a Mathematics Educator: Educational and Professional Background

I did not attend school from kindergarten to class two due to the *Nigerian or Biafran Civil War*. Given that both of my parents were educators, I had an early start before the war with informal education and home schooling that allowed me to leap right into class three. My school education began in January 1969 after my family except for my father (Jacob) relocated to Sierra-Leone as war refugees.

I received my primary education from class three to seven from the Fourah Bay College University Primary School (FBCU PS), Freetown, Sierra-Leone. FBCU PS tested me in various subject areas before making a determination that I was best suited for entry into class three. In fact, I was doing so well in class three that the principal, the late Mrs. Oluwole, had communicated to my mother (Europa) that if I made excellent grades in my final examinations, I would receive double promotion to class five. However, that did not happen because I accidentally broke my right hand swinging down the staircase railing at home. My right hand was in a cast for several weeks. I am right-handed. So, I could not take the final examinations to provide the evidence needed for double promotion to class five. Losing the opportunity of double promotion put a damper in my spirit. I did not put as much energy as I could have into my studies to earn it in subsequent years of class four and five until I got to class six by which time there was no more opportunity for this since class seven was the last year in primary school. I graduated from primary school in June 1973 at the age of 10.

I received my secondary education from form one to five from August 1973 to June 1978 at the Annie Walsh Memorial School (AWMS) in Freetown, Sierra-Leone. I spent two years from August 1978 to June 1980 in sixth form studying science from the Freetown Secondary School for Girls (FSSG), also in Freetown, Sierra-Leone. I was 15 years old when I finished from AWMS and 17 years old when I finished from FSSG.

AWMS was founded in 1948 by the Christian Missionary Society (CMS). It is the oldest girls' secondary school in Africa south of the Sahara. It is named after an English girl, Annie Walsh, whose dream was to become a missionary and educate children in Africa. She died tragically at age 20 on January 31, 1955, before she could begin to accomplish her dream. Her parents provided substantial funding to the CMS to memorialize the school in her name.

It is a family tradition on my maternal side for daughters in the family to attend the AWMS. My maternal grandmother, my mother and I attended this school. I have no daughter right now. I have emigrated to the United States. So it appears that my lineage will not be following in the footsteps of this AWMS tradition.

My mother (Europa) is a retired principal of the AWMS and a retired vice-principal of the FSSG. My immediate elder sister (Ikpendu) and I attended the FSSG. My sister attended the FSSG from form one to form five and I attended the FSSG for sixth form – lower and upper, since AWMS did not have a science program in the sixth form. To memorialize my mother's contributions to the AWMS, I established the Europa Wilson-Agwu Scholarship Fund for students who are exceling in mathematics. This fund is administered by the AWMS New York/New Jersey Alumni Association Incorporated, a 501c3 charitable organization.

The AWMS and the FSSG have nurtured and groomed many powerful and influential women. One of them noteworthy of mention in my NiWARD autobiography is my former English teacher at the AWMS, mentor, and friend, Dr. (Mrs.) Nemata Majeks-Walker. She was a good friend to me at a critical point when I was buying my home in the Bronx in New York, and has provided me with wise counsel on many professional and personal matters. "A friend in times of need is a friend indeed." She is doing phenomenal things for Sierra-Leonean women. She is the founder and 1st President of the 50/50 non-governmental organization in Sierra-Leone and a 2015 British Broadcasting Corporation (BBC) 100 Women Honoree.

The 50/50 organization is a non-partisan organization advocating and campaigning for increased political participation and equal representation of women in decision-making processes and initiatives at all levels in Sierra-Leone. It is the 1st African group to win the coveted Madeline Albright Award in 2007. A medium term goal of mine is to tell the mathematical stories of the women leaders of 50/50.

My father (Jacob) and I share some interesting name similarities or patterns related to our education even though he was not physically around to influence these for me. He attended primary school at Saint Joseph's Catholic School. I attended my confirmation classes at Saint Joseph's Convent Secondary School. Incidentally, Saint Joseph's Convent Secondary School is where I developed a fear of dogs. On one of the days I went for catechism class for confirmation, I was bitten on the right leg by a dog owned by a reverend sister in the convent. I had to receive a series of injections as a consequence of the dog bite. It is an experience that continues to haunt me till today. My father (Jacob) received his primary education through the benevolence of a Reverend Father Walsh. I received my secondary education at a school endowed with the name Walsh. He received his undergraduate education in regional geography from the Fourah Bay College University (FBCU). I received my primary education from the FBCU Primary School. It was as if my father's "invisible" hands were directing my educational steps (academic and religious) in Sierra-Leone while he was in Nigeria, determining where his own Nkechinyerem (His Own Gift From God) should be planted.

From 1980 to 1984, I attended the University of Nigeria, Nsukka (UNN). I majored in mathematics. I graduated in June 1984, at the age of 21, with a Bachelor of Science with Honors as the "Best Graduating Student" in the Department of Mathematics. My honors project is titled, *On the Stability of Solutions of Constant Co-efficient Second Order Equations and Systems*. My honors project advisor was the late Professor James Ezeilo. He was a founding father and the 1st Director of the Nigerian Mathematical Centre (NMC). He was a former Vice Chancellor of UNN. After graduation from UNN, I wanted to pursue a career in the Nigerian Army, Signals and Communications section, since I was interested in Cryptology. I could not do so as they were not accepting females at that time into sections of the army other than the medical corps section. I practice and teach Cryptology with my current scholarship in African culture and women's stories in STEM.

Incidentally, the NMC for which Professor Ezeilo was a founding father and 1st Director provided strong support to the Federal University of Technology, Akure (FUTA) for our Carnegie project through the current Director, Professor (Rev.) Adewale Solarin. The support of the NMC through Professor Solarin has continued by recruiting me as an African Diaspora Fellow to serve as a mentor for the 1st Pan-African Mathematics Olympiad for Girls (PAMOG) of the African Mathematical Union (AMU). I give thanks to God for the support of the present Director of the NMC (Professor Solarin).

Both the late Professor Ezeilo and the late Professor Isabella Ajaero, the only female faculty member of the Department of Mathematics at UNN while I was a student there, were instrumental in motivating me to pursue graduate education at University of Connecticut (UCONN), Storrs. Professor Ajaero wrote a strong positive personal letter of recommendation to her former dissertation advisor at UCONN, Professor Eugene Spiegel, on my behalf.

I was accepted into the master's degree program at UCONN, to begin in the Fall 1986 semester. I differed my admission to begin in the Spring 1987 semester, due to financial reasons. I landed in the United States via John F. Kennedy airport in Queens, New York City on January 14, 1987, in the midst of a terrible snow storm. This was my first experience with snow. I can remember how many times I slipped and fell that day because I was not wearing winter boots. I thank God that my falls did not result in any physical injury.

The Spring 1987 semester was my most academically challenging semester in the master's program at UCONN, since the pre-requisites for the courses I took that semester were only offered in the Fall semester and I was coming back to obtain a graduate degree after a two and a half year fallow period of studying mathematics in university at the undergraduate level. If I had not been an international student required by the United States immigration services to take only graduate courses due to admission into a graduate program, I would have been allowed by the Mathematics Department at UCONN to take undergraduate honors level mathematics courses in preparation for actually beginning my master's degree program in the Fall 1987 semester, as was their policy of accommodations provided to US nationals and non-international students admitted into their master's program in Spring semesters.

The semester of the Spring of 1987 was a critical test of my academic stamina and perseverance. I really struggled to succeed in my Numerical Analysis and Complex Analysis classes, staying late into the wee hours of the night studying while my colleagues were sleeping, trying to brush up on the pre-requisites that I had missed in the Fall 1986 semester by differing my admission to the Spring 1987 semester, and teaching myself how to program in the language of Pascal so that I could complete my Numerical Analysis programming assignments and projects. This semester yielded a completely different experience for me as a student who had always excelled in mathematics, never really having to dig my nose into my books where this subject was concerned throughout my school and undergraduate university educational life. Half-way through the semester, I almost quit the UCONN master's program. I was prepared to use my return ticket to go back home to Nigeria. The two key reasons that stopped me from doing so were the fact that I had resigned my position as a faculty member at Kaduna Polytechnic in Kaduna, Nigeria, and had no employment to go back to, and the fact that the late Professor Ajaero had demonstrated her faith in my mathematical abilities when she wrote a strong personal letter to her advisor, Professor Spiegel on my behalf. I acknowledged her faith in my mathematical abilities and struggled to make it through that semester. Afterwards, the remaining semesters of my master's program was smooth sailing for me. In January 1989, I graduated with a Masters in Science degree from UCONN. I majored in mathematics at the age of 25. I continued to pursue a doctoral degree in pure mathematics which I did not complete but earned 18 graduate credits. I transferred in 1990 from UCONN to Syracuse University (SU), Syracuse, to pursue a doctoral degree in mathematics education instead. Based on considerations of the employment market at that time and where I saw myself in the future, a doctoral degree in mathematics education as opposed to pure mathematics made better sense. UCONN did not have a doctoral degree program in mathematics education at that time. So a transfer to an institution that had a doctoral degree program in mathematics education was warranted. In August 1995, I graduated with a Doctor of Philosophy from SU. I was 32 years old, approximately two months away from my 33rd birthday. I majored in mathematics education with an emphasis in statistics and educational measurement. I also minored in curriculum and instruction with an emphasis in cultural foundations in the area of gender studies and multicultural education.

Prior to leaving Nigeria in pursuit of graduate education, I worked as a statistician for the Federal Office of Statistics (FOS), Zonal Office Eastern region, Enugu, Nigeria during my National Youth Service Corps (NYSC) under the directorship of Mr. Charles Essien, from August 1984 to July 1985. I was 22 years old at the end of my NYSC. I worked in the audit section doing spot checking of employment and other types of surveys conducted by the various state offices. This influenced my area of emphasis in my doctoral degree major in mathematics education. Upon completion of my NYSC, I worked as a Lecturer of Mathematics for Kaduna Polytechnic (KP), Kaduna, Nigeria, from 1985 - 1987. I was forced to quit my employment as a Lecturer at Kaduna Polytechnic when I was accepted to study mathematics, at the University of Connecticut (UCONN), Storrs, in their master of science degree program. This was due to institutional discriminatory practices in the nature of hiring Nigerians who were not from the Northern States, 10 states at that time. Nigerians from Southern states were hired like foreign contract workers rather than civil servants, even though Kaduna Polytechnic is indirectly supported by federal funds since states primarily derive their income from the Federal Government. If I were classified as a civil servant at Kaduna Polytechnic, I would have been able to secure paid or unpaid study leave and/or even sponsorship for my graduate education like my colleagues from the Northern States who left for further studies. Kaduna Polytechnic's loss of my employment was the ultimate gain of my current institution, Borough of Manhattan Community College (BMCC), City University of New York (CUNY) and my adopted land, United States. Not having a job to return to upon graduation from higher education I remained in the United States for practical training. I was employed by BMCC CUNY.

During the period of my graduate work in mathematics at UCONN, in addition to working as a graduate teaching assistant, I worked as Coordinator of the Mathematics Center responsible for the management and administration of the Center. At SU, I worked as a teaching fellow and as a teaching associate in the Mathematics Department. I received a few honors as SU such as the Future Professoriate Award, a Creative Dissertation Award, and the Chancellor's Meritorious Service Award for Student Leaders. I served in a number of student leadership capacities, which included service as a President of the African Students Union, as a President of the Association of International Students at Syracuse University, and as a Mathematics Department Graduate Student Organization Senator. My service in these capacities provided me with many valuable experiences in advocacy, leadership, and volunteer organizational management and administration. My best and worst experiences as a student in the United States occurred at SU, a story for my memoirs.

After acquiring my doctoral degree from SU in 1995, I wanted to come back home to Nigeria. My father (Jacob) advised against it because the employment market in Nigeria for a job in the university or some other job in my field in Ala Igbo (Igbo land) was bleak. So, I started looking for a job in the United States. My employment by BMCC CUNY in 1995 was once again the divine intervention of Jehovah-Raah (The Lord Is My Shepherd). It is a story for my memoirs at a later date. It suffices to say, that New York City (NYC) was never on my radar as a place to plant myself. However, in retrospect, God planted me firmly in the right place, where I could flourish, grow, and pursue my professional dreams. Once I came to this realization I began to branch out like a "cactus" tree planted in an oasis.

At BMCC CUNY, I was nominated as an Association of American Colleges and Universities Project Kaleidoscope Faculty for the 21st Century (AACU PKAL F21), Class of 1997. PKAL shaped me immensely as an advocate for STEM education and as a spokeswoman for equity for under-represented minority groups in STEM in the United States. This facilitated my professional growth. I served in many leadership capacities within PKAL.

At BMCC CUNY, I moved through the ranks from Assistant Professor to Associate Professor and finally to Full Professor, gaining tenure prior to becoming an Associate Professor and serving a three year term while I was a Full Professor as my college's Director of the Center for Excellence in Teaching, Learning and Scholarship (CETLS), formerly the Teaching Learning Center. I served in other elected and appointed leadership capacities which include service as a Deputy Chair of the Department of Mathematics, service as a Chair of the Instruction Committee and Curriculum Committee of the BMCC Faculty Senate, service as a BMCC Senator on the CUNY-wide Academic Senate, service on the Executive Board of the Professional Staff Congress (PSC) of CUNY, BMCC Chapter, and service as a Delegate-at-Large for PSC CUNY.

As Director of our college's CETLS, I chaired a 21 member faculty board and was responsible for coordinating and facilitating faculty development and scholarship. The CETLS underwent significant transformational growth under my leadership. I spearheaded the 15th year history celebration of the CETLS. This celebration brought name and date of service recognition of all Past Directors through a wall plaque at the CETLS.

My upward mobility at BMCC CUNY has not been like icing on a piece of cake. It is full of stories of hills and valleys, a story to be told after I retire from BMCC CUNY and have reached the biblical ripe age of three score and ten. However, I am Nma (Beautiful) Jacob, the overcomer. Overall, I have been richly blessed professionally by Jehovah-Raah. I cannot complain.

Chapter IV

4. Scholarship in African Culture and Women's Stories: My Research Journey

I am a historian of mathematics, an ethno-mathematician, a biographer, and a curriculum and assessment developer. My experience as a Visiting Minority Fellow at the Educational Testing Service in 2003 provided me with professional development in educational measurement and assessment. I support biographical works as a Life Patron and Life Fellow of the International Biographical Centre, Cambridgeshire, United Kingdom. I was a participant of the Institute of International Education (IIE) Fulbright CUNY Study Group Tour to China in 2004 on the *History of Chinese Mathematics and Mathematics Education in China*. My biographical focus is on Africans and African-Americans in science, technology, engineering, and mathematics (STEM) related fields.

From 1997 – 2002, I received professional development from the Mathematical Association of America, Institute in the History of Mathematics (MAA IHMT) on the history of mathematics, its uses in teaching and in developing historical curricular materials for teaching mathematics at the primary, secondary and tertiary levels. This included serving as Writing Team Chair for the development of the MAA historical modules in *Linear Equations* and *Polynomials* for the teaching and learning of mathematics and authoring biographical publications on two great African American pioneers and geniuses in mathematics, the late Dr. Ernest Wilkins Jr. and the late Dr. David Blackwell, whose lives inspire me and who I would like to emulate.

The late Dr. Ernest Wilkins Jr., was a nuclear scientist, mechanical engineer, and mathematician. He was the 1st African American to be inducted into the United States (US) Academy of Engineering. He contributed to the *Manhattan Project*. This was a project for creating the Atomic Bomb during the Second World War. It resulted in the end of the war when the United States dropped the Atomic bomb in Hiroshima and Nagasaki in Japan on August 6 and August 9, 1945, respectively.

The late Dr. David Blackwell was a mathematician and statistician. He was the 1st African American to be inducted into the US Academy of Sciences. He is one of the eponyms of the *Rao-Blackwell Theorem*, also known as the *Rao-Blackwell-Kolmogorov Theorem*. He was a pioneer in game theory and in writing textbooks in Bayesian Statistics. My work on his biography influenced the start of my research journey in African culture and women's stories with the Okwe (Mancala) game.

My professional development in these areas also took place through service for five years as the Historian of the American Mathematical Association of Two-Year Colleges (AMATYC), through service as a Writing Team Chair for the Chapter on Instruction for the signature document of AMATYC, *Beyond Crossroads*, and through service as a member of the *Beyond Crossroads Digital Products* Production Team. I was a 2000 pioneer recipient of the AMATYC INPUT Award for innovation in developing curriculum to teach statistics using biographical writing techniques, *Using a Threaded Discussion Web-based Software to Teach Statistics*.

I was a President of the American Association of University Women New York City Branch Incorporated (AAUW NYC Inc.). I served for two years as the Historian for the Branch and for two years as the Chair of its Writing Group. Under my presidency, I led the branch in the writing of its *She Touched Me* Series of biographies of members and honorees who have made a significant difference to the lives of others. The first two books in this series were submitted for the AAUW New York State biographical project on "women who have made a difference." This series had an immense impact on my scholarship as a women's story-teller.

My scholarship in the area of African culture and women stories in STEM began in 1998. I was awarded a New York Literacy Assistance Center mini-grant that made it possible for me to examine the teaching techniques inherent in the informal educational system of the Igbo culture, my culture. I did so under the mentorship of Reverend Fada, Dr. Jon Ukaegbu, a noted cultural anthropologist on Igbo symbolism and a Catholic Priest. My work in this regard, motivated further research interest. A mid-term goal of mine is to continue working with him on the Igbo calendar and the Igbo base four numeration system. We see evidence of this base four system in the capturing method used in the Okwe (Mancala) game.

Board games such as Okwe (Mancala) and Draughts (Checkers) are popularly played in Ala Igbo (Igbo land) and other parts of Nigeria. Okwe is a highly strategic game that can be used to reinforce many mathematical concepts in the area of game theory, number theory, probability, arithmetic and so on. It dates back to 5000 B.C. It was transmitted by slavery, commerce, and Islam. It is popularly known by its Arabic name, Mancala, which means to transfer and speaks about the action of transferring seeds from hole to hole in the one-to-one correspondence while playing the game.

My paternal grandmother (Omamma) was an expert Okwe (Mancala) game player. This humble market woman and 8th generation farmer from Umuapu, Agbakoli, Akoliufu, Alayi, who never had any formal education taught me how to play the game while I was an undergraduate student at UNN. We used to play together a lot during my vacation periods from UNN. With my strong mathematical background, I could never win her when we played. She could picture the tree diagram showing winning strategies for moves in the game, several moves ahead. I have developed mathematics curricular activities around this and I teach students in my BMCC Mat 150 – Introduction to Statistics and Mat 200 – Introduction to Discrete Mathematics Writing Intensive classes how to construct the decision tree diagram showing winning strategies of the game at some point in play and to compute probabilities related to the various paths of play.

God called my paternal grandmother (Omamma) away while I was pursuing graduate studies in the United States. Consequently, I was unable to attend the celebration of her life and burial.

God's Own: The Genesis of Mathematical Story-Telling

A few years later when I went home during the Christmas holidays to participate in the wedding of my brother (Uche), all that belonged to her had been taken by my siblings, cousins, and other people. By the divine intervention of Jehovah Shalom (The Lord is Peace), one item belonging to her was left that no one seemed to want. It was her Okwe (Mancala) game. To me, this was her most treasured possession. "The stone which the builders rejected was the cornerstone."

I guess the reason why her Okwe (Mancala) game was not taken is because it was meant for me to help me find peace at not being able to attend her burial and the celebration of her life and to have something of hers that I could pass unto my son (Ngozichukwuka) as an inheritance and wealth from his great grandmother (Omamma). I took possession of it as my own inheritance from my paternal grandmother (Omamma) and as a sign from her as to which path to follow in my scholarship pertaining to informal teaching techniques inherent in the Igbo culture. I continued my research journey in this direction with the Okwe (Mancala) game.

From 1999-2002, I was awarded three Professional Staff Congress (PSC), CUNY grants. These grants allowed me to travel to the ground zero point – Igboland in Nigeria, to conduct an in-depth examination of the Okwe (Mancala) game, and to engage in participant-observation interviews with expert players of the Okwe (Mancala) game, and with Eze's (traditional rulers) and Dibia's (traditional priests) with traditional knowledge about the history of the game, its motivations, and cultural symbolism. During the period between my 2nd and 3rd PSC CUNY grants, I was awarded a Louis Stokes Alliance for Minority Participation Research Initiative Participation (LSAMP RIP) Grant in 2001. This provided funding to support the development of curricular materials related to the Okwe (Mancala) game. These curricular materials are constantly being refined and expanded on. I have future plans to publish them with other mathematics curricular materials developed pertaining to African culture and women's stories in STEM.

Okwe (Mancala) is played in other Nigerian cultures, such as the Yoruba culture where it is called Ayo and has cultural motivations that differ from those of the Igbos. My medium term research goal is to examine the cultural differences in Okwe among different Nigerian cultural groups. By God's grace, I shall obtain the funding needed to accomplish this ethno-mathematics journey. Amen.

The end of the PSC CUNY and LSAMP RIP grant periods were followed with support through the BMCC CUNY Writing Across the Curriculum Program for developing Writing Intensive (WI) courses in Mat 200 – Discrete Mathematics and Mat 150 – Introduction to Statistics. This brought about an examination of African women in STEM and rural African women's work interwoven into the teaching and learning of these courses. To facilitate my research in African culture and women's stories in STEM, I mentored students to engage in smaller research projects within my scholarship area rather than in other areas through the Collegiate Science and Technology Entry Program (CSTEP) and the Louis Stokes Alliance for Minority Participation (LSAMP) Program.

During this same period, I was a 2004 pioneer recipient of the newly established CUNY Community College Collaborative Incentive Research Grant that I helped to advocate for in 2003 within the Community College Committee of the CUNY-wide Academic Senate. I was awarded this grant for the project, *An Investigation Into the Patterns of Uses and Effects of Self-Medication in Caribbean Immigrant Communities*. My Co-principal Investigators for this project on self-medication were Professor Brahmedeo Drewprashad from the Science Department (Chemistry faculty) and Professor Barbara Tacinelli from the Nursing Department, both from BMCC CUNY. We also involved BMCC CUNY students in research internships and engaged other faculty members from the Mathematics, Science and Nursing departments to work with us on this project. My work on this project triggered my interest in examining mathematical connections related to medicinal herbs commonly used in Africa, such as *Moringa* (*The Miracle Plant*) and also other traditional plants and flowers. I recognized that since plants and flowers are all around us it was a great way to promote class equity in the teaching and learning of mathematics for rural and lower-income schools who cannot afford expensive resources.

Finally came the Eureka momentum, a Carnegie African Diaspora Fellowship for the project, *Culture and Women's Stories: A Framework for Capacity Building in Science, Technology, Engineering and Mathematics (STEM) Related Fields,* to spend time again at the ground zero point – Nigeria, to immerse myself in ethno-mathematics research and co-curricular development activities at CEGIST, FUTA. By engaging in this project the journey of my scholarship in mathematical story-telling of members of the Nigerian Women in Agricultural Research and Development (NiWARD) began.

Chapter V

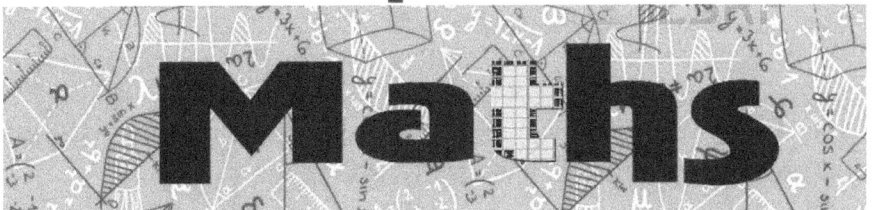

5. **The Carnegie African Diaspora Fellowship Project – *Culture and Women's Stories: A Framework for Capacity Building in Science, Technology, Engineering and Mathematics (STEM) Related Fields***

This project was one of the round one 31 projects selected for funding in 2014 under the new African Diaspora Fellowship initiative of the Carnegie Foundation. This initiative is coordinated by the Institute of International Education (IIE). I am one of the 33 pioneer fellows funded under this initiative. I received funding to partner and collaborate with colleagues at the Federal University of Technology, Akure (FUTA), under the aegis of its Centre for Gender Issues in Science and Technology (CEGIST) with leadership on the FUTA end by the late Dr. (Mrs.) Mojisola Edema, former Acting Director of CEGIST, Dr. (Mrs.) Olayinka Ogunsuyi, then Assistant Director of CEGIST and Ms. Olabuunola Willaims, Coordinator of the Nigerian Women in Agricultural Research for Development (NiWARD) Program at CEGIST at that time.

Our project goals were to develop culturally based and gender sensitive curricular materials; to engage in teaching-research on their use; and to prepare Nigerian educators in using them to teach mathematics and related disciplines at the primary; secondary and tertiary level. The aim of this project is to foster student innovation and creativity that is linked to the science and technology of their culture; to nurture and mentor girls to consider science, technology, engineering and mathematics (STEM) related careers; and to provide poor and rural communities with inexpensive curricular resources that are easily obtainable from the local community.

We worked steadfastly towards accomplishing these goals with crucial off-site support from the Pan-African Strategy and Policy Research Group (PANAFSTRAG) through their Maths, Science, Technology, to Innovate Indigenous Knowledge Systems (MSTIIKS) Program under the leadership of Mrs. Arinola Bello, the PANAFSTRAG Program Coordinator, and under the leadership of Professor Emeritus Stella Williams, Executive Board member of PANAFSTRAG. An expository article of our work on this project is published by the Mathematics Teaching Research Journal Online at:

http://www.hostos.cuny.edu/MTRJ/archives/volume7/issue2/volume7issue2full.pdf.

As a testimony of the success of our project, I was one of the five fellows selected by the Carnegie Foundation to present our project and my experience as a Carnegie African Diaspora Fellow (CADF) at the 57th annual African Studies Association (ASA) conference CADF Roundtable on November 22, 2014. Our project and I were featured under impact stories "STEM and Women's Stories: US-Nigeria Academic Collaboration Advances Women in STEM" by the Institute for International Education in its 2014 annual report, see:

http://www.iie.org/en/Who-We-Are/Annual-Report/Impact-Stories/Carnegie-African-Diaspora-Program.

Dr. Nkechi Madonna Agwu
Carnegie Africa Diaspora Fellowship

Based on the outcomes of our project, I was selected by IIE to represent the Carnegie African Diaspora Fellowship Program as a presenter on the panel "The Potential of Diaspora Engagement: Trade, Skills Development, Education and Youth Mentorship in Science, Technology, Engineering and Mathematics" at the June 10, 2015, World Bank forum in Washington D.C., USA, on "Harnessing the Diaspora for Promoting New Directions in Trade, Education & Skills for Workforce Development". I gave a presentation that discussed our project and my related experiences; that discussed my thoughts related to diaspora engagement and capacity building in STEM; and that discussed my thoughts related to empowerment of women and girls.

I gave a presentation on my work in this regard at the inception meeting on *Gender Mainstreaming, Agric-Business Innovation and Fostering Entrepreneurship Investment From the National Agricultural Research Systems*, held at the Agricultural Research House in Mabushi, Abuja, on September 22, 2015. My presentation titled, *Mathematical Stories of NiWARD, Ndebele Dolls, Vertex-Edge Graphs, Number Patterns and Cryptology* was well received by the participants, stimulating great discussion and feedback. The picture below is of some of the participants at my presentation with me at the Agricultural Research House.

Work on this project will continue in 2016 at the Nigerian Mathematical Centre (NMC) in Abuja, Nigeria, and at BMCC CUNY.

A few curricular activities that illustrate the outcomes of this project and the intended work related to NiWARD are given in at the end of my story. A long term goal or vision of mine is the publication (online or hard-copy or soft-copy print or video or other types of multi-media) of several modules of curricular activities in different STEM fields and even non-STEM fields on African culture and women's stories that expand upon what is being done in the Nigerian context with NiWARD to other groups of Nigerian women in STEM and also to Global African women in STEM. I have faith in Jehovah Jireh (The Lord Will Provide) that He shall bring to the NiWARD table the right funders, stakeholders, and leadership team to make this a reality. He has brought the project this far. The best place to be is in the hands of The Absolute Infinite (God). I have placed this project which demonstrates His existence and mathematical form in His creative hands.

Chapter VI

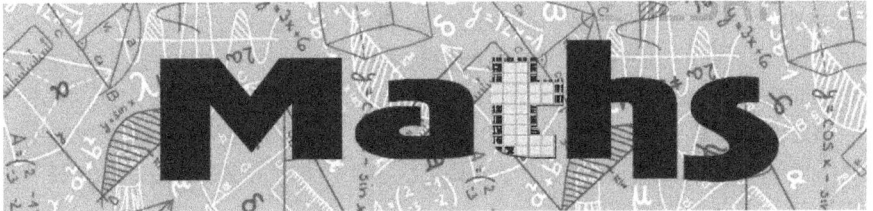

6. Personal Challenges: Coping Strategies

My life testifies of God's love. I am blessed by God with a successor. He is my teenage son - Ngozichukwuka (The Blessing Of God Is The Greatest) Jacob (Successor or Overcomer) Ayemere (Imortalize My Name) Durosimi (Wait And Bury Me) Agwu (Igbo patron deity of health and divination employed to explain good and evil), aka ngozi1 or bondock9 - the kid with amazing skills (his YouTube movie aliases). Like his grandfather (Jacob), he has a creative mind. He is frequently at work conducting various types of scientific experiments and investigations in my home (his laboratory).

According to the Igbo naming tradition, the names of my son (Ngozichukwuka) shall speak to my life experiences and express my hopes and expectations of him. He shall live longer than me to succeed me. He shall overcome all challenges. By his successful accomplishments, he shall tell the story of Nkechi (God's Own) to the world to immortalize my name as Nma (Beautiful) Jacob. His life is meant to give God all the glory and testify of His many blessings. One of such blessings is that I did not suffer from any of the deadly complications that come with being gestational diabetic.

My family (immediate and extended), over many generations, have been plagued by Diabetes and have died from its complications. My mother (Europa) and I have been diagnosed with it for several years. One of my biological sisters was gestational diabetic like me but a more severe case. She was on insulin. Even my late husband (Nicholas) was on insulin for his Diabetes. He died of its complications.

This Diabetes pattern shall not be a co-traveler and roommate with my son (Ngozichukwuka) and his lineage by God's divine healing powers. Neither shall it take control over my life by attacking my vision, my feet, my kidneys, my pancreas, my heart, my nerves, my bones, my teeth, my gums, and my skin. Nor shall it create urologic and digestive problems and challenges of sexual dysfunction and dementia for me. By His stripes I am healed. Amen! This disease that took many lives on my maternal ancestral line, shall not take my life in the mighty name of Jesus, according to His word in Luke 13:12. Amen.

Jehovah Rapha (The Lord That Heals) has filled my son (Ngozichukwuka) with a good genealogical understanding of how this generational disease has plagued our family. He watches his diet and even cautions me about watching mine. Recently, he even did as his high school science research project a study on Diabetes and related preventive health care maintenance.

Vertex-edge tree diagram mappings of a person's genealogy are extremely important in preventive medicine. It can help a person to understand and educate themselves about potential hereditary diseases commonly suffered from by Global African citizens, such as Diabetes, Sickle Cell, Cancer, and Hypertension. Incidentally, these are all hereditary diseases in my family lineage either on the maternal or paternal side. My father (Jacob) and my paternal grandmother (Omamma) were diagnosed as Hypertensive. My mother was diagnosed with Breast Cancer in her senior years. Both of my parents (Jacob and Europa) are carriers of the Sickle Cell trait. Jehovah Rapha has broken this Sickler trait for my lineage, being born neither a Sickler nor a carrier of the trait. Abba Father, no type of Cancer shall take possession of my body according to your word in Psalm 121. Amen. Jehovah Rapha (The Lord That Heals) please take control. These diseases will not be transferred to my son (Ngozichuwkuka) and his lineage, in the mighty name of Jesus. Amen.

Vertex-edge tree diagram mapping can help a person begin early to properly manage their health to prevent the occurrence of such hereditary diseases in their life or minimize their complications. Understanding vertex-edge tree diagram mapping is also important for contact tracing in Epidemiology.

One of the reasons why Nigeria which is one of the most populous countries in Africa was able to contain Ebola when it was brought into Lagos, the most populous city in Nigeria, by the Liberian-American Patrick Sawyer, was by doing an effective job with complete contact tracing or mapping of the vertex-edge tree diagram of this first Ebola patient. So, teaching our children as an early age about genealogy or about the mathematics of contact tracing will facilitate public health education. This is something that is extremely important in under-developed countries, like my homeland Nigeria, with a high poverty rate that impacts the ability of its citizens to receive proper or adequate health care when they fall ill.

Right from my womb my son (Ngozichukwuka) chose a similar path like me to be a uniquely defined. Remember I walked the path of deviation in the spelling of my name as Adeleine. My water broke on my birthday of Thursday, October 8, 1998, around 8:30 a.m. in the morning. I thought he was going to be my birthday gift. However, he chose to deviate – not to share the same birthday with me. Unlike me, he was not in a hurry to begin his life trajectory as Jacob, my successor. I was in labor for over 23 long, extremely painful, but joyous hours. I have never laid on my back for as long as I did giving birth to him. He was born the next day, at 8:05 a.m. on Friday, October 9, 1998. He was born by Caesarian Section because the delivery doctor said he was in distress. I was operated on three times related to his birth. The 1st operation was to deliver him. The 2nd operation was the same day not too long after his delivery because the medical team, thank God, discovered that they left a swab in my womb after sewing me up. Consequently, I suffered from infection for several days after his delivery. The 3rd operation was a year later as a consequence of scarring of my womb from these two surgeries.

My son (Ngozichukwuka) is indeed Jacob, the overcomer. He is a survivor. He is my miracle child! He overcame distress at birth, even smiling to me at the operating table when they were taking him away to the nursery after delivery, essentially telling me that God was watching over me for the second operation and I would be fine. He overcame hearing and speech problems as a child. This caused him to be held back a year in Kindergarten and to use amplifiers in school to aid his hearing. This resulted in teasing by his classmates. Now my son (Ngozichukwuka) is producing movies and communicating fluently. He is using amplifiers no longer for hearing but for movie production.

God's Own: The Genesis of Mathematical Story-Telling

In addition to the reversal of hearing loss, my son (Ngozichukwuka) overcame a terrible knife gang mugging in June 2013 on the day he was celebrating his middle school graduation with classmates. He overcame critical illnesses that caused him to be hospitalized and could have left him dead while we were visiting Nigeria in 2007. This is what God can do! God's grace is sufficient to see you through any challenges, 2 Corinthians 12:9.

Metaphorically speaking, I have overcome several personal challenges related to vision, hearing, speech and my other senses that left me lying on the road for dead. Like God did for Jacob in Psalm 44:4, He delivered me from the jaws of death, wiped me clean, gave me a new name, and give me a chance to re-build and restore myself. He did the same like-wise for me professionally. He took the stone which the builders rejected and made it the cornerstone. What I have accomplished has been not by my power but by His grace upon my life and the angels He used to guide me and support me through the process. Given all the mountains He removed to get me to the place I am today, I know that "nothing is impossible with God" Luke 1:37. This is my personal mantra.

When I run into a road block related to professional and personal matters, I seek the face of God for "Open Heavens" according to his word in Psalm 23, in the mighty name of Jesus. Amen. I know He has a divine purpose for my life for I have literally escaped death like a cat with nine lives. Some of these escapes relate to car accidents in which the car was totaled but I was not severely injured. When I write my memoirs at a later date when I am past three score and ten, I shall share the full details of the challenges I speak about in metaphorical terms. By then the work of God in my life would have come full circle, one complete revolution or 360 degrees or 2π radians. For now, I would like to say this to those who read my story here to always "keep hope alive" and to never lose sight of your dreams no matter what challenges you face and how severe they are.

Chapter VII

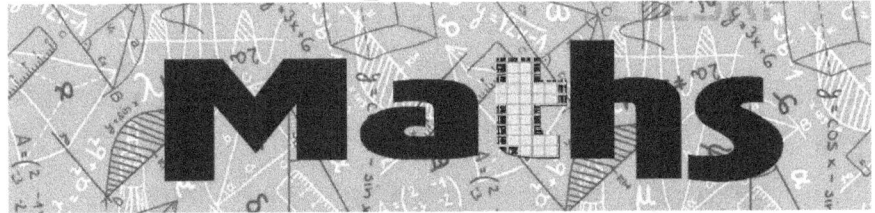

7. *The Absolute Infinite*: My Philosophy

Like George Cantor, *The Father of Set Theory*, I am interested in how mathematics reveals the existence of God. I see myself as a servant of God or God's own (Nkechi), similar to Cantor. Like him, I have been endowed with Nkechinyerem (My Own Gift From God). My gift is a talent for African story-telling using vertex-edge graphs and number patterns reflected in a person's life.

Like George Cantor, I think of God as The Absolute Infinite. Cantor communicated his thinking about God in his writings linked to his ground breaking work on infinite sets and transfinite numbers. My scholarship in African culture and women's stories in science, technology, engineering and mathematics (STEM) related fields is innovative because it connects the teaching and learning of mathematics to story-telling about African culture and women in STEM using vertex-edge graphs and number patterns as the framework.

The Absolute Infinite (God) is at the center of my ground-breaking work on the mathematical story-telling of the Nigerian Women in Agricultural Research for Development (NiWARD). He inspires my creative mind. He gives me the vision to make the necessary mathematical connections to our African culture, to the indigenous mathematical knowledge evident in the work of rural women, and to the stories of NiWARD. He directs my pen on what to write, the words, sentences, and paragraphs to use.

The Absolute Infinite (God) is present in the "infinity" circular ponds in a small inner enclosed courtyard by the entrance to the Coronation Hall at the Deji's Palace in Akure Kingdom, Ondo State, Nigeria. Deji is the name for the Monarch of Akure.

Figure: The "Infinity" Circular Ponds at the Palace of the Deji of Akure Kingdom

Akure is where the Federal University of Technology, Akure (FUTA) is located. It is the work of The Absolute Infinite (God) that ensured that during the short period I was planted for our Carnegie Project at FUTA, I was granted permission by the Regent, Her Royal Highness Princess Adetutu Adesida, for a visit to the Deji's Palace. The Absolute Infinite (God) needed me to see His existence in the "infinity" circular ponds. He opened the doors on the visitation date of Friday, August 22, 2014, for me to be granted permission by the High Priest, Chief Adebayo, the SAO of Akure Kingdom, a retired educator, for entry into the Palace, a guided study tour of the Palace, and participant observation interviews of the tour guides. The SAO has the final decision over the Regent on entry into the Palace by visitors. High Priest, Chief Adebayo, told me that I was wearing the garment of Africa after interviewing me to decide whether or not to grant me entry into the Palace. He was implicitly communicating that I have a significant role to play in bringing the indigenous mathematical knowledge of African people to the fore-front of the history of mathematics and its uses in teaching.

Crown Prince (Engineer) Adesina, escorted us in our guided historical tour of the palace which included taking pictures, some of which were of the triangular symbolism in the Coronation Hall. I thank God for the enthusiasm of Crown Prince Adesida. He provided me with rich informational print materials about the history of Akure Kingdom and the genealogy of its Dejis. I am using this material to develop mathematics curricular activities on Akure Kingdom.

According to my historical tour guides, these "infinity" circular ponds can never be filled even though there is no source of leakage at the bottom or anywhere for water to seep through or other explainable reasons that relate to seepage or evaporation. This filling process yields infinite sequences whose sums to any given number of terms are supposed to be strictly increasing, thus changing by virtue of growth. Yet in reality based on what these tour guides communicated they remain constant and equal to the volume of the ponds. This to some may be considered a contradiction or an impossible event. To others like me it is a demonstration of the existence of The Absolute Infinite (God). Remember "nothing is impossible with God", Luke 1: 37. He, The Absolute Infinite, is the same God that took a constant of five loaves and two fishes and multiplied it exponentially to feed the multitudes of over five thousand men, and many more women and children, with 12 baskets of left overs, Matthew 14:17 -21. He is the God who through Moses parted the Red Sea, creating a wall of water, for the Israelites to pass through and escape slaughter by the Egyptian army, Exodus 13:17. He is the God who created me, His own (Nkechi) with an assignment to reflect His mathematical properties through mathematical story-telling related to African culture and women in STEM. The beauty and grace of His mathematical form as The Absolute Infinite is magnified in my mathematical genome.

Chapter VIII

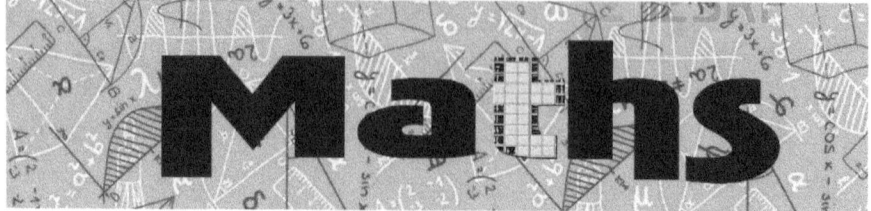

8. My Mathematical Genome: A Tale of Vertex-edges Graphs and Number Patterns

INkechi (God's Own) am many times an *Equilateral Triangle*. This is a three-sided polygon with all sides equal and all angles equal to 60 degrees. It is a vertex-edge graph with three sides known as a three sided odd cycle or complete graph. Both types of vertex-edge graphs have the number three as their Chromatic number since each vertex is connected to the other two.

As an advocate and spokeswoman for equity, justice and recognition for Global African women in Science, Technology, Engineering and Mathematics (STEM) and for nurturing the African girl child to consider STEM related careers, my three connected vertices are equity, justice, and recognition. In my scholarship promoting gender, class and cultural equity for the African continent in the history of mathematics and its uses for teaching and learning mathematics, my three connected vertices are gender equity, class equity, and cultural equity for Africa. These two isomorphic and similar triangles for my scholarship related to the Nigerian Women in Agricultural Research for Development (NiWARD) are immersed one on top of the other (inner and outer concentric triangles). Try graphing them as concentric triangles.

These two concentric remind me of the triangular symbols on the columns that serve as the cornerstone in the Coronation Hall at the Deji's Palace in Akure Kingdom, Ondo State, Nigeria (see picture below). My immediate interest is finding out if this triangular symbolism in the Coronation Hall of the Deji's Palace is connected to the role that a soon to be crowned Deji must play in the Kingdom and his or her responsibilities or is it just purely decorative?

Figure: Triangular symbolism in the Coronation Hall of the Palace of the Deji of Akure Kingdom.

There are other types of vertex-edge graphs, viz., paths, wheels, trees, cycles with more than three sides, complete graphs with more than three sides in my mathematical genome. Some easy paths you can identify in my life are: the path of my formal education from primary school to highest educational level (FBCU PS – AWMS – FSSG – UNN – UCONN – SU) and the path of my employment (FOS – KP – UCONN – SU – BMCC CUNY).

Vertex-Edge Graph: Path of Employment

Since a path always has Chromatic Number two, both paths can be colored with a minimum of two colors so that neighboring vertices have a different color. For both paths, the degree of the two end vertices is one because they are only connected to one vertex, either at the left or at the right, and the degree of each of the inner vertices are two because they are connected to two other vertices, one at the left and the other at the right. The path of my employment has five vertices, the two end vertices (FOS and BMCC CUNY). The path of my formal education has six vertices, the two end vertices (FBCU PS and SU). Your assignment is to draw the path of my formal education and color it with its Chromatic Number of two colors so that neighboring vertices have a different color.

One easy cycle with more than three vertices that you can identify in my life is the cycle of the countries that I moved around from and to as a consequence of the Nigerian or Biafran Civil War (Nigeria – Fernando Po – Liberia – Sierra-Leone – Nigeria). This cycle has four vertices. Therefore it is an even cycle with Chromatic Number two and the degree of each vertex is two since each vertex is connected to another vertex on the left and on the right.

Vertex- Edge Graph: Cycle of Countries of Relocation Due to the Nigerian or Biafran Civil War

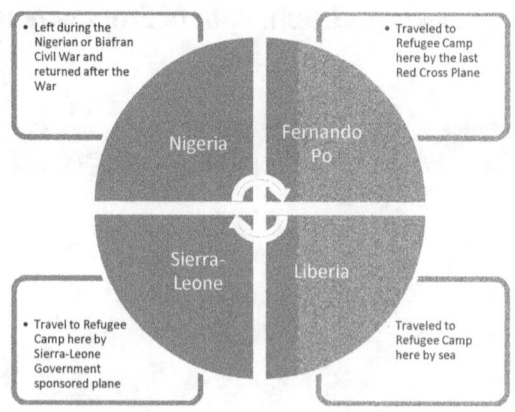

We can also identify a few wheels with triangles around the central hub in my life. This type of wheel is also a complete vertex-edge graph with four vertices and is pyramidal. It has Chromatic Number four because every vertex is connected to each other. There are three isomorphic wheels characterizing the similarities between my parents and myself in being mathematics teachers, educational pioneers and mentors of many children and youth. These similar characteristics of (mathematics teacher, educational pioneer, and mentor of many children and youth) form the vertices of the triangle around the central hub. My father (Jacob) is the central hub for the first wheel, my mother (Europa) is the central hub for the second wheel, and I (Nkechi) am the central hub for the third wheel. Connecting the hub of these three wheels gives the triangle of my birth connection with vertices father (Jacob), mother (Europa) and child (Nkechi). Represented in three dimensions, these three wheels are three pyramids from my story with trilateral base. See whether you can draw these three wheels showing the triangular connections of their central hubs?

Culturally, Ndi Igbo (Igbo people) have used pyramidal symbolism in ancient times based on the archeological evidence of the *Nsude Pyramids* found in Agbaja, Kogi State or in Udi, Enugu State, Nigeria. The Igbo patron deity Ala or Uto was believed to reside at the top. Ala is the female deity of the earth, morality, fertility, and creativity. My significant church affiliations rest on a tripod with its legs being Holy Ghost Cathedral (for protective covering of my birth records), Vineyard International Christian Ministries (for harvesting my creativity) and Glorious Miracle Embassy International (for glorifying, praising and giving thanks for the fertility of my creativity). Resting at the top of this tripod is The Absolute Infinite (God of earth, morality, fertility and creativity), the God of Sarah, Elizabeth and Hannah in the Bible, similar to the Nsude Pyramids. Try graphing this tripod.

I have quite a few genealogy trees. Some of them are my family farming genealogy tree, my family AWMS genealogy tree, my family Jacob name genealogy tree, my family Adeline name genealogy tree, my family Nma name genealogy tree, my family early widowhood genealogy tree, my family Diabetes, Sickle Cell, Cancer, and Hypertension genealogy trees, and my research mentor genealogy tree. Try graphing one of these genealogy trees to determine the spanning sequence of the branches.

The triangle is by far the most dominant vertex-edge graph in my life. There are triangles in my names. There are triangles in my baptism. There are triangles in my war experiences. There are triangles in my educational and professional paths. There are triangles in my research journey. There are triangles in my genealogy of various forms. There are triangles in the nature of my scholarship in African culture and women's stories. There are triangles in my personal challenges, coping strategies, and philosophy. There are triangles in my NiWARD persona.

The figure below provides a graphical representation of a few triangles in my name. Earlier on, I identified two triangles in my scholarship in African culture and women's stories. See whether you can identify some of the triangles in my baptism, war experiences, educational and professional path, research journey, genealogy, personal challenges, coping strategies and philosophy , and in my NiWARD persona.

Triangles in my Name

1. Jacob's triangle #1: Name of maternal grandfather, father and son
2. Jacob's triangle #2: Name of father, last name of alias, name of son
3. Nkechi's triangle: Nkechi (God's Own), Nkechinyere (Gift of God), Nkechinyerem (My own gift from God)
4. trinity triangle in feminine form: Madonna (Mother of God), Nkechi (God's own child), Agwu (Spirit form of God)

Figure: Four Triangles in my Name

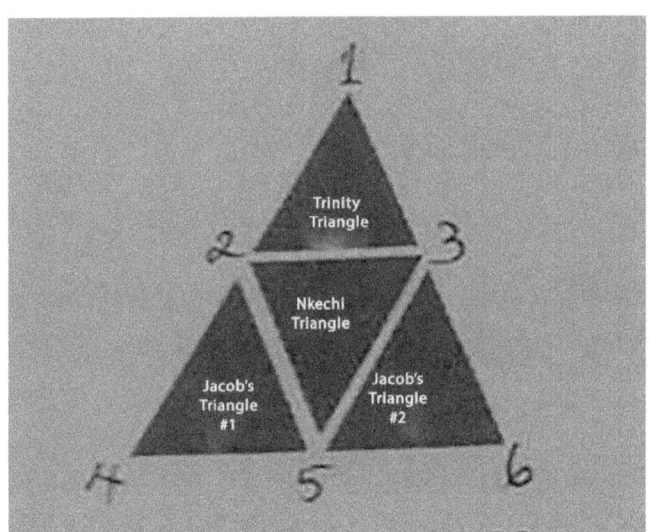

These triangular arrangements, some of them hidden and some of them as clear as daylight, are everywhere in my life from birth to present. Therefore, I am an infinite sequence of *Triangular Numbers*, 1, 3, 6, 10, 15, 21, 28, 36, 45,... The third term 6 in my triangular sequence symbolizes the triangular figure above that shows four triangles in my name with six vertices. My triangular sequence is infinite because I am Nma Jacob blessed by God with succession in my lineage which shall continue till infinity. You can find my triangular sequence in the third diagonal of the *Pascal's Triangle*. Prove by induction that the sum of my triangular sequence for any given number of terms n is the tetrahedral number given by $[n(n+1)(n+2)]/6$.

My associated series is 1, 4, 10, 20, 35, 56, 84, 120, 165,... You can find my associated series in the fourth diagonal of the Pascal's Triangle. My associated sequence of successive differences for my sequence of triangular numbers is 2, 3, 4, 5, 6, 7, 8, 9,... You can find my associated sequence of successive differences in the second diagonal of the Pascal's Triangle.

Pascal's Triangle: First Ten Rows

1									
1	1								
1	2	1							
1	3	3	1						
1	4	6	4	1					
1	5	10	10	5	1				
1	6	15	20	15	6	1			
1	7	21	35	35	21	7	1		
1	8	28	56	70	56	28	8	1	
1	9	36	84	126	126	84	36	9	1

Sum of row one is $2^0 = 1$
Sum of row two is $2^1 = 2$
Sum of row three is $2^2 = 4$
Sum of row four is $2^3 = 8$
...
Sum of row n is 2^{n-1} for any natural number n. This is a common exponential function which has many real life applications in genetics, business, computer science, and so on.

You now have the important task of identifying as many of these triangles as you can find in my story and determining what term they are in my triangular sequence. Will you be able to find all of them? No, because I am an infinite triangular sequence even though I am countable or discrete.

In base 10, you can show that the digital root of any term in my triangular sequence is a member of the set {1, 3, 6, 9}, with a periodic sequential pattern of 1, 3, 6, 1, 6, 3, 1, 9, 9 that has period of nine. Therefore, every number in my triangular sequence is either divisible by three or has one as a remainder when it is divided by nine.

A digital root also known as the repeated digital sum of a natural number is the single digit value which is obtained by an iterative process of finding the sum of the digits on each iteration by using results from the previous iteration to compute a digital sum. The process continues until a single digit number is obtained. The number of iterative steps is known as the number's additive persistence. Therefore:

- $1 = 9 \times 0 + 1$; Digital root of $1 = 1$; Additive persistence of $1 = 0$
- $3 = 9 \times 0 + 3$; Digital root of $3 = 3$; Additive persistence of $3 = 0$
- $6 = 9 \times 0 + 6$; Digital root of $6 = 6$; Additive persistence of $6 = 0$
- $10 = 9 \times 1 + 1$; Digital root of $10 = 1 + 0 = 1$; Additive persistence of $10 = 1$
- $15 = 9 \times 1 + 6$; Digital root of $15 = 1 + 5 = 6$; Additive persistence of $15 = 1$
- $21 = 9 \times 2 + 3$; Digital root of $21 = 2 + 1 = 3$; Additive persistence of $21 = 1$
- $28 = 9 \times 3 + 1$; Digital root of $28 = 2 + 8 = 10 = 1 + 0 = 1$; Additive persistence of $28 = 2$
- $36 = 9 \times 4$; Digital root of $36 = 3 + 6 = 9$; Additive persistence of $36 = 1$
- $45 = 9 \times 5$; Digital root of $45 = 4 + 5 = 9$; Additive persistence of $45 = 1$

There are other meaningful sequences in my life, such as the sequence of my numerological number patterns. I have developed a sample accompanying arithmetic curricular activity for finding some of the numerological number patterns in a person's life. Use this activity to find the numerological number patterns in my life.

Another meaningful sequence in my life is the increasing sequence reflecting some significant numbers in my NiWARD story, viz., 3, 4, 5, 6, 8, 9, 10, 11, 12, 16, 18, 22, 26, 33, 2936. A sequence is strictly increasing if the successive term is greater than the preceding term.

- 3 is for the class in which my formal education started, for the number of agricultural engineering patents of my father (Jacob) and for the number of vertices in the triangle which is my dominant vertex-edge graph

- 4 is for my baptism day, for my doctoral defense day, for the age my father (Jacob) lost his father, for the day we landed in Sierra-Leone as refugees, and for the number of vertices of the wheels in my life that I have highlighted

- 5 is for my age when the Nigerian or Biafran Civil War started and for being one of the five selected pioneer Carnegie African Diaspora Fellows to present their project and experiences at the 57th annual African Studies Association conference

- 6 is for the age my mother (Europa) lost her father, for the number of children born by my paternal grandmother (Omamma) and for the number in my triangular sequence that represents the four triangles in my name

- 8 is for my birthday, one of my two significant months and my paternal grandmother being an 8th generation farmer. This significant number is another triangle in my life

- 9 is for the birthday of son (Ngozichukwuka), for my father being a 9th generation father and for the month of the terrorist attacks on the World Trade Center

- 10 is for one of my significant months, for me being a 10th generation farmer, for me being born in the 10th month of the year and for the day and month I graduated as a trained chaplain licensed to practice anywhere in the United States and around the world

- 11 is for my son (Ngozichukwuka) being of 11th generation and for the day of the terrorist attacks on the World Trade Center

- 12 is for the disciples in my Mat 200 – Discrete Mathematics class at BMCC CUNY, Spring 2015 semester who are collaborating with me to begin the vertex-edge graph story-telling of NiWARD

- 16 is for the birthday of my father (Jacob) and the birthday of my alias (Nma Jacob)

- 18 is for the age in months when my father (Jacob) identified my son as having a creative mind and for how many months prior to the birth of my father (Jacob) that his maternal grandfather died

- 22 is for the day High Priest, Chief Adebayo, the SAO of Akure Kingdom told me I was wearing the garment of Africa, for the day I gave the presentation at ASA, for how old I was when I started working at the Federal Office of Statistics, Enugu, upon graduation from the University of Nigeria, Nsukka and for the hour of my birth (10 p.m. or 22 hours) on October 8, 1962

- 26 is for the day my Eureka moment was triggered by the late Dr. (Mrs.) Mojisola Edema

- 33 is for being one of the 33 pioneer Carnegie African Diaspora Fellows and for my age at the end of the year of my graduation with my doctorate degree

- 2936 is for my baptismal record number that placed my birth

This finite sequence of significant numbers has 15 terms. The digital root for 15 is 6 and it has persistence of 1 since $15 = 1+5 = 6$ and the process of reduction ended after the first sum. The sum of all its terms is $3119 = 3+4+5+6+8+9+10+11+12+16+18+22+26+33+2936$ which also has a digital root of 5 and additive persistence of 2, since $3+1+1+9 = 14 = 1+4 = 5$ and the process of reduction ended after the second sum. If we represent this particular finite sequence of significant numbers in its digital root form it becomes 3, 4, 5, 6, 8, 9, 1, 2, 3, 7, 9, 4, 8, 6, 2, since the digital root of each of the single digit terms of 3. 4, 5, 6, 8, 9 are themselves and:

- $10 = 1+0 = 1$ based on digital root computation
- $11 = 1+1 = 2$ based on digital root computation
- $12 = 1+2 = 3$ based on digital root computation
- $16 = 1+6 = 7$ based on digital root computation
- $18 = 1+8 = 9$ based on digital root computation
- $22 = 2+2 = 4$ based on digital root computation
- $26 = 2+6 = 8$ based on digital root computation
- $33 = 3+3 = 6$ based on digital root computation

God's Own: The Genesis of Mathematical Story-Telling

- $2936 = 2+9+3+6 = 20 = 2+0 = 2$ based on digital root computation

This sequence of digital roots has a sum of 68 = 3+4+5+6+8+9+1+2+3+7+4+8+6+2. This leads to a digital root of 5, since 68 = 6+8 = 14 = 1+4 = 5 with a persistence number of 2 since we carried out the reduction process twice. Try repeating this exercise by using a different increasing sequence that you construct from other significant numbers in my story.

Now you have analyzed me in my vertex-edge graphical and number sequence forms, your next important assignment is sculpturing me as a beautiful Ndebele Doll representation of *Our Lady of Africa*. Our Lady of Africa is a bronze image of the Immaculate Conception considered as a consolation of the afflicted and a protector of all Africans of all faiths, Christians (Catholic and non-Catholic) and Moslems alike. Remember, I am a Black Madonna. My Ndebele doll representation should be crowned with fertility yams because I am planting seeds in an uncharted creative realm with faith in The Absolute Infinite (God) for birthing a great harvest.

My Ndebele Doll representation should be dressed in a beautifully woven Akwete cloth filled with symmetrical and asymmetrical patterns of equilateral triangles, cycles, wheels, trees, and complete graphs, sequenced periodically. It should have a Talking Drum and Shekere musical instrument interspersed with strings of cowrie and palm nut shells planted around its waist like *Mgbaji or Lagidigba or Jigida or Loo or Nkwa-Isin* (waist beads) to signify my transformation into my beautiful, graceful and noble triangular NiWARD persona.

The palm nut shells symbolize my work in agricultural research and development, planted in me by my father (Jacob). The cowrie shells symbolize the wealth of my harvest which shall multiply like the popular exponential function with base two (with real-world applications in genetics, business, computer science, and so on) whose terms are the sums of the rows of the Pascal's Triangle.

The Talking Drum should be decorated with nine strings tied around it to represent the period of the periodic sequence 1, 3, 6, 1, 1, 3, 1, 9, 9,…, of the digital roots of the terms of my triangular infinite sequence and my Life Path Number. Leaves from the Moringa Tree (The Miracle Plant) should be wrapped around these nine strings to symbolize healing blessings of Jehovah Rapha (The Lord That Heals) in my life.

The Shekere musical instrument should be decorated with 15 cycles (circles) of threaded coral beads. The number of beads in each cycle should be equal to a number in the sequence of digital roots 3, 4, 5, 6, 8, 9, 1, 2, 3, 7, 9, 4, 8, 6, 2 for my significant numbers.

You should wear around the neck of my Ndebele Doll representation, two coral beaded necklaces with six coral beads each to represent the digital root of the number of terms in the sequence of my significant numbers and the digital root of the sum of the terms in my significant numbers and to represent the two rows and six holes in an Okwe (Mancala) game board. These coral beaded necklaces are to communicate my nobility, given that in many traditional Nigerian cultures, monarchs, queens, crown princes, and princesses, wear sacred coral beaded necklaces, often in pairs to signify nobility.

In the words of Galileo Galilei, the famous Italian astronomer, physicist, mathematician, engineer and philosopher who played a major role in the scientific revolution during the Renaissance period, "the universe cannot be read until we have learnt the language and become familiar with the characters with which it is written. It is written in the mathematical language and the letters are triangles, circles and other geometrical figures, without which it is humanly impossible to comprehend." We see these letters in my mathematical genome and in the triangular symbolism and infinity circular ponds at the Palace of the Deji of Akure.

In awe and wonder at my nobility, beauty and grace which you can see clearly through my Ndebele Doll representation, I have magnetically drawn you into my triangular maze and my universe. You are temporarily lost in my maze at a significant intersection of crossroads. You have the important task of finding the right path, the NiWARD path. This is the path that will lead you out of the maze into Jacob's House in Alayi.

Nma Jacob is your light out of the maze. Close your eyes and envision my NiWARD Talking Drum, Shekere musical instrument and Akwete cloth. You will find your way out of the triangular maze but now transformed into a NiWARD disciple having learnt the language and become familiar with the characters of my universe, just like biblical Saul of Tarsus was transformed to Apostle Paul.

As a disciple, you have the responsibility to tell others the story of NiWARD using your own Nkechinyerem (Your Own Gift From God) curricular activities which you received in the maze. Remain forever blessed and inspired to go out and be the creative being that God molded in like form, using this project to prove His existence as The Absolute Infinite. He is alive in you!

Chapter IX

9. The Curriculum Developer in Action: Sample NiWARD Curricular Activities

The activities given below can be adapted for any collection of African women in science, technology, engineering and mathematics (STEM) related fields or any collection of African women in leadership. It can also be adapted for any collection of women or men, irrespective of their ethnicity or professional endeavors.

Activity I: Ndebele Dolls - Graph Theory

Background

The Ndebele people are an ethnic group from Southern Africa. Their entire life style (architecture, clothing, artifacts, dolls, jewelry, and so on) is deeply rooted in basic geometry, number patterns, graphs, and other forms of symbolism. Like much of the mathematical traditions and indigenous knowledge of African people south of the Sahara, you will hardly read about the Ndebele people in mathematics text books that teach concepts of geometry, number theory, and graph theory.

Traditionally, the Ndebele people use dolls in symbolic ways. Two examples are given as follows. When an Ndebele woman wants a child, she will make a doll and name the doll. When she eventually bears the child, she will name the child after the doll. When an Ndebele man wants to propose to a woman to marry her, instead of giving her an engagement ring as a sign of courtship, he gives her a doll made by him.

Ndebele Dolls are created out of recycled materials or waste products – "do what you can with what you have from where you are". In every culture you will find dolls. So this activity can be modified and adapted for dolls from other cultural groups, such as the Igbos, Yorubas, Hausas, Edos, Fulanis, Ijaws, Tivs, and so on in Nigeria or for dolls from any culture in the world. However, the choice of Ndebele Dolls for this activity is symbolic because Ndebele people are highly mathematical in their way of life. Birthing and naming of a new being and marriage to person is connected to the creation of a doll. Also, they nurture the creative mind-set and environmental protection through doll making from recycled waste products.

How to make an Ndebele Doll

Materials needed

1. Recycled empty water bottle for creating the body of the doll.
2. Recycled clothing for the costume and head-gear of the doll.
3. Sand, stones, rice, beans, beads, coins or other items to stuff the bottle to give the doll weight. You can also use water.
4. Clay, model magic or other suitable items to create the head and hair of the doll.
5. Colored beads, buttons and other embellishments to create the face of the doll and decorate its costume.
6. Fabric or wood glue or any other type of strong glue to stick the costume on the body of the doll and the head unto the body of the doll.
7. Scissors to cut clothing for the costume of the doll and other materials that need cutting.
8. Wire hangers or sticks or other similar material to create arms and legs for the doll (if you wish).

Directions

1. Give the body of the doll weight, by stuffing the water bottle with one or more of the materials identified in #3 under materials needed.
2. Put glue all around the water bottle.
3. Cut clothing that will be sufficient to wrap around the water bottle and glue the clothing unto the bottle. For a regular 1.25 pint or 20 fl.oz. water bottle, a 12 inch by 12 inch or 1 foot by 1 foot cloth will suffice.

4. Use your clay or model magic or other suitable items to create the head of the doll and put it on top on the water bottle by sticking or gluing it on or sewing in on or any other means that will keep it permanently fixed to the top of the bottle.
5. Use your colored beads, buttons or other embellishments to create the face of the doll and decorate the body of the doll.
6. Use wire or sticks to pierce through the body of the doll to create arms and legs (if you wish)

For a more hands-on experience of how to create the doll, watch the African Views Organization ACE YouTube video on how to make an Ndebele doll at

http://www.youtube.com/watch?v=HamUbtroHcA.

Activity

Using the directions given for how to construct Ndebele dolls:

1. Construct an Ndebele doll to represent a woman from the group of NiWARD whose biography inspires you and whose footprints you would like to follow.
2. Name your Ndebele doll after this woman and bring your doll to exhibit.
3. Decorate your doll to reflect three vertex-edge graphs you see in the woman's story, and identify the name (path, cycle, wheel, complete graph, triangular arrangement, tree, and so on) of the two vertex-edge graphs if possible.

4. Color the three vertex-edge graphs for their Chromatic Number, state their Chromatic Number, and explain why you cannot color them with a number of colors less than what you claim is the Chromatic Number.
5. Use counting techniques and principles to count and state the number of vertices and edges of the two vertex-edge graphs on your doll and create a table showing the vertex-edge graph, its name, its Chromatic Number, and the number of its vertices, edges and the degree of each vertex.
6. Develop an accompanying power-point presentation that tells a story of the life of this woman. The story should present the three vertex-edge graphs you reflected on your doll representation of this woman, their Chromatic Number, and the number of vertices, edges and degrees of each vertex for each of the graphs. The story should explain why you selected those three vertex-edge graphs over the other types that are evident in the biography of the woman. It should also discuss aspects of the woman's life that inspires you and that you would like to emulate.

Activity II: Numerology – Arithmetic, Ciphers and Sequences

Background

Numerology is a belief in the divine mystical relationship between a number and some coinciding events. It comprises the art of recognizing relationships between symbols (numbers and letters, names, and dates) and what they represent to gain a deeper understanding of the universe or of a person. This belief system exists in many world cultures, including Nigerian cultures such as the Igbo culture with the Afa system and the Yoruba culture with the Ifa system. Unfortunately, due to globalization, Nigerians are losing their knowledge of the Afa and Ifa numerological systems.

The ancient Pythagoreans practiced a form of Numerology organized by their leader the Greek philosopher and mathematician Pythagoras who is credited as having discovered the famous *Pythagorean Theorem* for right angled triangles that states that the square of the length of the Hypotenuse of a right angled triangle is equal to the sum of the square of the lengths of the two other sides.

The Pythagorean system combined mathematical disciplines of the Arabic, Druid, Phoenician, Egyptian and Essene sciences. It is known today as the western form of Numerology. It is the most popular and commonly practiced form of Numerology. Our world is becoming more computerized and dependent on numerical systems. So, this ancient spiritual science is becoming highly significant today. You can learn more about Numerology and become a part of the pattern project at

http://www.numerology.com.

You can also find more details here about the interpretations of the different numerological numbers.

Activity: Find Numerological Number Patterns of a woman from the group of NiWARD and explain what each numerological number signifies.

1. Write the name of a woman from the group of NiWARD woman as a finite number sequence using the Pythagorean Cipher and state the number of terms of this sequence.
2. Find the five core numbers, or as many of the five as you can, if the required information is provided in her biography or can be found: (a) Life Path Number (b) Expression Number (c) Personality Number (d) Heart's Desire Number (e) Birthday Number. You can find detailed procedures for determining these numbers together with illustrations at

http://www.numerology.com.

3. Represent these five numbers sequentially in the order of (a) – (e) given above.
4. Find the Sun Number of this woman, if the required information is provided in her biography or can be found.
5. Find the Hidden Passion Number of this woman, if the required information is provided in their biography or can be found.
6. Repeat #1 using the well-known Caesar Cipher which is a shift of the alphabet three places to the right.
7. Repeat #1 using an alphabetic Cipher of another non-western cultural group or one created by a person from a non-western culture.

Algorithm: For Computing the Life Path Number

The procedure for computing the life path number of a person is as follows: reduce the month, day, and year of birth to a single digit or one of the master numbers – 11, 22 or 33 by continuously summing the digits if they are two or more (like you did when finding digital roots); then add the sum of the reductions for month, day and year, and reduce this sum to a single digit by the same procedure used before if it is not a single digit or a Master Number. The detailed procedure and the meaning of each life path number are given at

http://www.numerology.com.

Example: Life Path Number for Nkechi Madonna Adeleine Agwu, NiWARD volunteer

Step-by-step computation

1. Month: My birth month is October which is the 10 month. 10 reduces to $1+0 = 1$
2. Day: My birthday is the 8th. 8 is a single digit so no reduction is necessary
3. Year: My birth year is 1962. 1962 reduces to $1+9+6+2 = 18$. 18 reduces to $1+8 = 9$
4. The sum of the reductions for month, day and year is $1+8+9 = 18$. 18 reduces to $1+8 = 9$
5. Therefore, my numerological Life Path Number is 9. This number signifies someone with global awareness. It signifies a person that is helpful, compassionate, aristocratic, sophisticated, charitable, generous, humanitarian, romantic, cooperative, self-sufficient, proud and self-sacrificing.

Pythagorean Cipher: For Finding the Expression, Personality and Heart's Desire Numbers

1	2	3	4	5	6	7	8	9
A	B	C	D	E	F	G	H	I
J	K	L	M	N	O	P	Q	R
S	T	U	V	W	X	Y	Z	

Numerical Values for the Letters in Nkechinyere Madonna Adeline Agwu (Birth Name) Using the Pythagorean Cipher

N	K	E	C	H	I	N	Y	E	R	E
5	2	5	3	8	9	5	7	5	9	5

Therefore, the Pythagorean Cipher number sequence for Nkechinyere is: 5, 2, 5, 3, 8, 9, 5, 7, 5, 9, 5. This finite sequence has 11 terms.

M	A	D	O	N	N	A
4	1	4	6	5	5	1

Therefore, the Pythagorean Cipher number sequence for Madonna is: 4, 1, 4, 6, 5, 5, 1. This finite sequence has seven terms.

A	D	E	L	I	N	E
1	4	5	3	9	5	5

Therefore, the Pythagorean Cipher number sequence for Adeline is: 1, 4, 5, 3, 9, 5, 5. This finite sequence has seven terms.

A	G	W	U
1	7	5	3

Therefore, the Pythagorean Cipher number sequence for Agwu is: 1, 7, 5, 3. This finite sequence has four terms.

So, the Pythagorean Cipher number sequence for my birth name Nkechinyere Madonna Adeline Agwu is: 5, 2, 5, 3, 8, 9, 5, 7, 5, 9, 5, 4, 1, 4, 6, 5, 5, 1, 1, 4, 5, 3, 9, 5, 5, 1, 7, 5, 3. This finite sequence has 29 terms, the sum of the number of terms of the individual finite number sequence for each of my names (11+7+7+4 = 29).

Algorithm: For Computing the Expression Number

The procedure for finding a person's Expression Number is as follows: write out the full name the person was given at birth, first, middle and last names. Use the Pythagorean Cipher above to obtain the numerical value for each of the letters. Add the numbers of the person's first name and reduce them it to a single digit, using the procedure for computing digital roots. Repeat this procedure with the person's middle(s) and last names. Finally, add the three single-digit numbers together and reduce them to another single-digit number to reveal your Expression number. If you encounter a Master Number - 11, 22, 33 in the reduction process do not reduce it any further. Interestingly, you can find these three master numbers reflected in certain aspects of my life story.

Example: Expression Number for Nkechi Madonna Adeleine Agwu, NiWARD volunteer

Reminder – use should be working with the person's birth name
1. Nkechinyere: $5+2+5+3+8+9+5+7+5+9+5 = 63 = 6+3 = 9$
2. Madonna: $4+1+4+6+5+5+1 = 26 = 2+6 = 8$
3. Adeline: $1+4+5+3+9+5+5 = 32 = 3+2 = 5$
4. Agwu: $1+7+5+3 = 16 = 1+6 = 7$
5. Sum of the single-digit numbers; $9+8+5+7 = 29 = 2+9 = 11$. This is a master number so we do not reduce it.
6. Therefore my Expression Number is 11. This number signifies someone with a powerful presence that possesses a bridge between the conscious and unconscious. It signifies a person with deep faith and with psychic abilities.

Algorithm: For Computing the Personality Number

The procedure for finding a person's Personality Number is as follows: write out the full name the person was given at birth, first, middle and last names. Use the Pythagorean Cipher above to obtain the numerical value for each of the letters. Sum up the numbers corresponding to the consonants in the person's name. Reduce the numbers to a single-digit using the digital root computation process. If you encounter a Master Number – 11, 22, 33 in the reduction process do not reduce it.

Example: Personality Number for Nkechi Madonna Adeleine Agwu, NiWARD volunteer

My Personality Number is 6. Use the algorithm given above for computing the Personality Number to verify this. Reminder – you should be working with the person's birth name. The Personality Number of a 6 signifies a person who is a caretaker. It signifies a person who is responsible, loving, self-sacrificing, protective, sympathetic and compassionate.

Algorithm: For Computing the Heart's Desire Number

The procedure for finding a person's Heart's Desire Number is as follows: write out the full name the person was given at birth, first, middle and last names. Use the cipher code above to obtain the numerical value for each of the letters. Sum up the numbers corresponding to the vowels in the person's name. Reduce the numbers to a single-digit using the digital root computation process. If you encounter a Master Number – 11, 22, 33, in the reduction process, do not reduce it.

Example: Heart's Desire Number for Nkechi Madonna Adeleine Agwu, NiWARD volunteer

My Heart's Desire Number is 11. Use the algorithm given above for computing the Heart's Desire Number to verify this. Reminder – you should be working with the person's birth name. The characteristics of 11 were previously highlighted under the section for Personality Number since 11 is my Personality Number.

Example: Birthday Number for Nkechi Madonna Adeleine Agwu, NiWARD volunteer

There is no algorithm for a person's Birthday Number as it is just the day of the month the person was born. My birthday is the 8th. The number 8 signifies balance and power. It signifies a person with a talent for business. It signifies a person who approaches issues with originality creativity. It signifies a person who is confident and ambitious and always ready for a challenge.

Example: Core Numbers for Nkechi Madonna Adeleine Agwu, NiWARD volunteer

Therefore, the sequence of my five Core Numbers is: 9 – Life Path, 11 – Expression, 6 – Personality, 11 – Heart's Desire, 8 – Birthday Number.

In Numerology, what do these Core Numbers represent?

1. The Life Path Number is the blue print for your life.
2. The Expression Number is about your life lessons and how you handle them.
3. The Personality Number is about the first impression people have of you.
4. The Heart's Desire Number is about what you really want in life and the rationale behind your actions.
5. The Birthday Number is about a special talent you have that will be elevated at some point in your life.

Algorithm: For Computing the Sun Number

The procedure for computing the Sun Number of a person is as follows: add the person's month and day of birth. Reduce the sum to a single digit using the process of digital root computation, Master Numbers – 11, 22 or 33 should also be reduced. A detailed procedure for this computation with illustrations and the meaning of each Sun Number are given at:

http://www.numerology.com.

Example: Sun Number for Nkechi Madonna Adeleine Agwu, NiWARD volunteer

Step-by-step computation

1. Month: My birth month is October which is the 10 month.
2. Day: My birthday is the 8th.
3. Sum of month and day = 10 + 8 = 18
4. Compute digital root of sum of month and day = 1+8 = 9.
5. Therefore, my Sun Number is 9. This number signifies someone who is an idealist.

Algorithm: For Computing the Hidden Passion Number

The procedure for finding a person's Hidden Passion Number is as follows: write out the full name the person was given at birth, first, middle and last names. Use the Pythagorean Cipher above to obtain the numerical value for each of the letters. Identify the modal (most frequently occurring) number in the letters of your name. This number is your Hidden Passion Number. If you have multiple modes, it means you have multiple Hidden Passion Numbers.

Example: Hidden Passion Number for Nkechi Madonna Adeleine Agwu, NiWARD volunteer

My Hidden Passion Number is 5. It is has the highest occurrence with a frequency of 11 for the numbers in the Pythagorean Cipher. Use the algorithm given above for computing the Hidden Passion Number to verify this. A Hidden Passion Number of 5 signifies a person who is an adventurer. It signifies a highly adaptable and versatile person with a talent for language and writing.

Project I - Creative Writing: The Mathematical Genome of a Woman of NiWARD - Vertex-Edge Graphs and Number Patterns in Their Life

1. Read the biography of a woman from the group of NiWARD whose life inspires you or who you would like to emulate.
2. Engage in Activity I and II related to her story.
3. Now you have read her biography and engaged in these activities, you are uniquely positioned to construct her mathematical genome, craft her tapestry and sculpt her as an Ndebele doll. In the African story-telling tradition, tell us about the mathematical genome of this woman in NiWARD who inspires you or who you would like to emulate.

For an illustration of this type of mathematical story-telling, see the section seven on my mathematical genome.

Chapter X

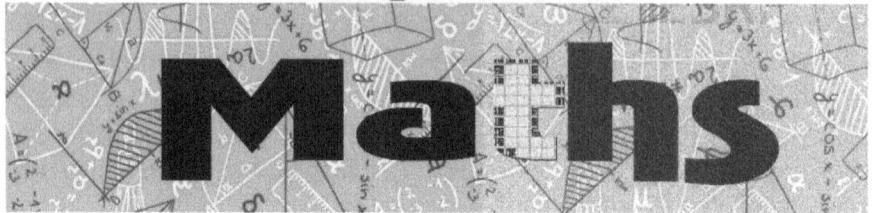

10. *The Triangular Arrangement of Nma Jacob* and Related Curricular Activities

This chapter is dedicated to my Borough of Manhattan Community College (BMCC), City University of New York (CUNY) students of MAT 200: Introduction to Discrete Mathematics, Writing Intensive courses during the academic years of 2013 -2015. This was a crucial period for my self-development as a mathematical story-teller of women in STEM.

The Triangular Arrangement of Nma Jacob

My name is Nkechi Madonna Adeleine Agwu
Sometimes called NMA by my loved ones
N stands for Nkechi
M stands for Madonna
A stands for Adeleine
I am beautiful
NMA means beautiful in Igbo, my mother tongue
My names form the triangular arrangement of NMA

I am the daughter of Jacob Ukeje Agwu
He was a 9th generation farmer
From the royal home of Ndi Mbada in Agbakoli Alayi
I am the 10th generation of Umu Apu
Apu was a great warrior from Akwa Ibom
He had a lion for a pet
In Igbo culture, it is traditional to answer your father's first name as your last name.
So, my name is Nma Jacob

I am a triangular arrangement
An odd cycle and a complete graph with three vertices
I symbolize the trinity – three in one
In Igbo, Nkechi means God's own child
In Igbo, Agwu means God, the spirit of healing and divination
Agwu is known to take on a feminine form
In Italian, Madonna means Mary, Mother of God
So, I am the feminine form of the Trinity

I am Nma Jacob
Jacob is a faithful servant of Chukwu
Chukwu means God in Igbo
Nma means beautiful in Igbo
Adeleine means noble woman in German
So, I am the beautiful faithful servant of Chukwu
Of noble heritage from the royal household of Ndi Mbada
My characteristics of beautiful, faithful and noble form a triangular arrangement

~ Adapted from the Collection of Poems of Nma Jacob

Curricular Activities Related to the Triangular Arrangement of Nma Jacob

1. Graph the three triangles of Nma Jacob as concentric rings of similar equilateral triangles stacked on top of each other?

2. What type of solid pyramidal figure have you created?

3. Find the volume of the pyramidal figure that you have created?

4. Sculpt the pyramidal figure using modeling clay.

5. Sculpt an Ndebele doll of Nma Jacob.

6. Create an Okwe (Mancala) game using an empty dozen egg carton and 48 seeds.

God's Own: The Genesis of Mathematical Story-Telling

7. Paste the Ndebele doll of Nma Jacob to the top of the pyramidal figure to culturally symbolize the Nsude pyramids in Agbaja or Udi with Ala or Uto, the female deity of the earth, morality, fertility and creativity at the top.

8. Paste your Okwe game to the base of your pyramidal figure to culturally symbolize farming and harvesting through the sowing of seeds into holes and capturing of seeds into the storehouses.

Conclusion

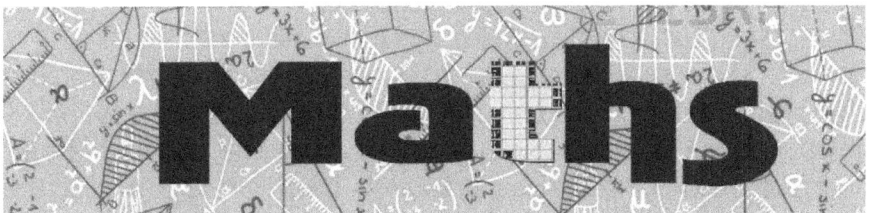

I n conclusion, I end with a quote from the Association of American Colleges and Universities (AAC&U) November 2012 working conference program booklet. "- To address global challenges – food , health, energy, security – the world needs both liberally educated STEM professionals and scientifically and quantitatively literate college graduates in all fields equipped with integrative knowledge, inquiry and analytical skills, innovation, vision, and inspiring leadership." However, we are pumping our African women and children from STEM because we fail to invest adequately in and support programs or projects, such as my Carnegie Project, that will provide our African women and children with access, opportunities, and avenues to excel in STEM; or that will nurture, groom and mentor them to become productive citizens; or that will tell the stories of African women leaders in STEM like the late Dr. (Mrs.) Edema; or that will tell the stories of rural African women like my grandmother (Omamma) whose work reflects indigenous mathematics, scientific and technological knowledge. A genius lost to our society through this pump is one too many for he or she is someone who could do great things for our nation just like Madam C. J. Walker, the first African American female millionaire, political activist and philanthropist whose faith in God and story inspires me through the challenges she overcame or just like Dr. Robert Moses, the civil rights leader and founder of the Algebra Project. The Algebra Project is a national mathematics literacy movement serving thousands of inner-city and rural public schools in the United States. This project, which I was fortunate to be involved in as a New York City professional development consultant and trainer, significantly influenced my thinking about the human rights perspective of mathematical literacy. Are we going to leave our African women and children behind in our efforts to meet this global challenge or are we going to go the extra mile and take the necessary steps to ensure that we have a sufficient and diverse pool of African women and children in STEM, being mindful of the fact that women are half the sky!

God's Own: The Genesis of Mathematical Story-Telling

I am a Nigerian, Sierra-Leonean, African-American widowed woman, who is a survivor of the civil wars or Nigeria, Sierra-Leone and the 911 World Trade Center terrorist attacks. The trajectory of my life includes experiences of displacement, homelessness, living in refugee camps, single-parenting of a child with hearing and speech needs and many other issues that from all indications could have set me up for failure. However, despite the odds against me which included being the sole black female student in the graduate mathematics programs of the University of Connecticut (UCONN), Storrs, and Syracuse University (SU), Syracuse, while I was a student there, I was fortunate to have access, opportunities, and avenues to excel in mathematics and other areas relevant to public service in STEM, through mentors such as the late Dr. William Brazziel, a former Dean of the School of Education at UCONN, Dr. Howard Johnson, a former Associate Vice-Chancellor and Dean of the Graduate School at SU, Dr. Joan Burstyn, a former Dean of the School of Education at SU. Dr. Howard Johnson was a financial savior to me in the last summer of my student days at Syracuse University enabling the completion of my dissertation. Dr. Burstyn rekindled my love and passion for history and poetry being a historian and poet herself. She was highly instrumental in removing a stumbling block that might have caused me to drop-out from the doctoral program during the last semester of my dissertation process. As a professional, I also had mentors like Dr. Sadie Bragg, the 1st African American President of the American Mathematical Association of Two Year Colleges and a former Provost and Vice President of Academic Affairs at BMCC CUNY, and Dr. Victor Katz, a historian of mathematics and one of the Directors of the Mathematical Association of America Institute in the History of Mathematics and it's Uses in Teaching, that helped to open professional doors for me.

My mentors took a vested interest in me. They encouraged me to avail myself of a variety of programs or projects that nurtured and groomed me to become a grassroots community leader. They showed me how mentoring can foster success. I am following their footsteps at BMCC CUNY as a mentor to many students in the various research programs and as a faculty advisor for over 15 years to the African Students Association.

Today, I am blessed as a Professor of Mathematics at the Borough of Manhattan Community College (BMCC), City University of New York (CUNY) a minority-serving institution that recognized my leadership potential and invested in me early in my career as a college STEM faculty member. I am empowered to give back to my communities through my employment at BMCC CUNY, and through my active involvement in many civic societies and professional organizations that have prepared me as a leader, such as the Black Women for Black Girls Giving Circle (BWBGGC), the American Association of University Women (AAUW), the American Mathematical Association of Two-Year Colleges (AMATYC), Mathematical Association of America (MAA), Project Kaleidoscope of the Association of American Colleges and Universities (PKAL AAC&U), Bronx Volunteer Fire Department (BVFD) Company #4, Vineyard International Christian Ministries (VICM) Inc., Anglican Church of the Pentecost, Worldwide Association of Small Churches, African Peoples Alliance, and United Peoples Inclusive, among others. I advocate and work within these organizations for the empowerment of women and children, with a special focus on groups that are under-represented in STEM.

I have many stories about my life to tell that will nurture, groom, and prepare young women and children for a career in STEM. This story of mine, *God's Own: A Genesis of Mathematical Story-Telling* is just one of the many. It will be followed by a second part of curricular activities framed around my mathematical genome and cultural interests.

If this story captivates one young woman or child to consider a career in STEM, through that person shall come exponential growth. It shall come within the exponential function with base two that is part of my mathematical genome as previously illustrated.

After reading this story of mine, bless a young woman or child interested in a career in STEM by giving them a copy of my story to read. Seat back and now watch for exponential growth like the five loaves and two fishes that Jesus Christ used to feed over 5000 men, women and children with 12 baskets of left-overs. I declare in the mighty name of *The Absolute Infinite* (God), that this anointed autobiography *God's Own: A Genesis of Mathematical Story-Telling* shall feed multitudes exponentially! Remain forever blessed.

God's Own: The Genesis of Mathematical Story-Telling

References

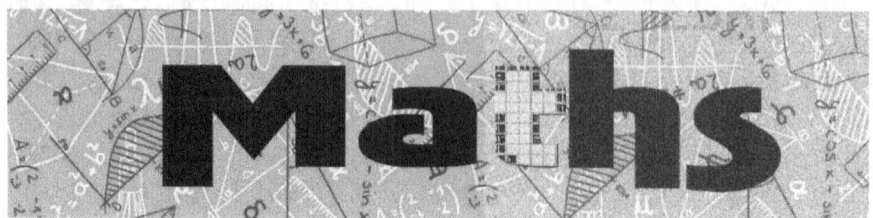

Gerdes, P., *Women, Art and Geometry in Southern Africa*, Africa World Press, Inc., Trenton, NJ, 1998.

Lumpkin, B., "Africa in the Mainstream of Mathematics History," In Ethnomathematics *Challenging Eurocentrism in Mathematics Education* by Powell, B., and Frankenstein M. (Eds.), 1997, pp. 101-117.

https://www.youtube.com/watch?v=dkycEMQ0fyg

http://www.hostos.cuny.edu/MTRJ/archives/volume7/issue2/volume7issue2full.pdf

https://www.youtube.com/watch?v=CdWar62RrwE

http://www.hostos.cuny.edu/MTRJ/archives/volume7/issue2/volume7issue2full.pdf

http://www.iie.org/en/Who-We-Are/Annual-Report/Impact-Stories/Carnegie-African-Diaspora-Program

http://www.youtube.com/watch?v=HamUbtroHcA

http://www.numerology.com

Reviews

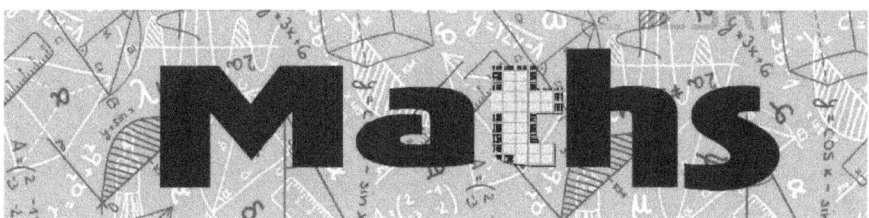

There are seven reviews in this section. The choice of seven is to symbolize completeness, perfection, and the creative timeline of God as reflected in the book of Genesis, chapters 1 and 2 in the Bible. Represented symbolically by two reviews from women and five reviews from men is the miracle of exponential growth in the five loaves and two fishes that feed over 5000 men, women and children with 12 baskets of leftovers. It is not by accident that the reviews by the women are the two fishes and the reviews by the men are the five loaves. Fishes have the ability to birth other fishes whilst loaves do not have the ability to birth other loaves. The Absolute Infinite, the God of Sarah, Elizabeth, and Hannah, who changed their stories from barrenness to fruitfulness, shall change my story by breaking the chains of any barren areas in my life and shall bring froth life to all who read this story in the mighty name of Jesus. Amen.

The Two Fishes

In this book, the author, a "warrior woman" with an indomitable spirit and a Mathematician par excellence, adequately demonstrates that the experience of losing significant stories of her life prior to a war in the country of her birth – Nigeria is a contributing factor to her passion for story-telling today and her love of history and genealogical mapping.

Dr. Nemata Majeks-Walker
Founder and 1st President of 50/50

This book is heartwarming and inspiring. It is a must read for all people of faith.

Bishop (Dr.) Ebony Kirkland
Presiding Prelate of Worldwide Association of Small Churches
& Church of the Living God Worldwide

The Five Loaves

Vineyard International Christian Ministries
140 Teller Avenue, Bronx, New York 10456
Tel.: (718) 538-9211 (Voice); (718) 542-7417 (Fax)
E-Mail: Vineyardusa@aol.com

Dr. Nkechi Agwu is from first to last a woman of her time, place, and culture: an African American, an insider everywhere. Her upbringing, education, achievements, or rather accomplishments all by the Grace of God, as well as, her academic and intellectual positions bring her in contact with individuals whose names appear in many aspects of this autobiography. One glaring fact among many as refreshing as this treatise may appear, pen and ink are too weak to capture the totality of who this Professor Agwu is and any mere or ordinary view of her person is frequently very different from the real portrayal of her everyday life.

The journey of Dr. Nkechi Agwu takes her from the dangerously political Nigeria of the 1960's to an academic haven of Sierra Leone Judea and cosmopolitan institution of Fourah Bay College University, where the intellectual storehouse of the early West African coexists uneasily with epicurean delights, hedonistic abandon, sharp business dealing natural resource, as well as, minerals, and later on the ever-present threats of political manipulation, personal deceit, and ethnic quarreling as in all other West African nations.

Opening the autobiography of Dr. Nkechi Agwu is like stepping into a fabulous bazaar in a familiar land, where every turn might lead to a familiar pleasure or a refresh relish. Dr. Agwu's story is both intensely personal and deeply vernacular in presentation. Her excitements, delights, and disappointments are achingly but in faith human. At the same time, her experiences foreshadow the wrenching changes that most of us that grew up in this same cultural region will experience in the catastrophic diaspora that we will continue to find ourselves in the coming years.

This autobiography of Nkechi Agwu, *God's Own: The Genesis of Mathematical Story-Telling*, is an exciting and engrossing epic that will richly reward the reader. It deserves its glorious place among the other equal successes in the African-literary works.

Bishop (Dr.) Joe Omeokwe
Founder and General Overseer of Vineyard International Christian Ministries
Professor Emeritus and Past Chairman of the Department of Mathematics,
Touro College Professor of Mathematics, Boricua College
Archdeacon of the Anglican Missionary Diocese of the Trinity, North East region

---

One may wonder why Dr. Nkechi Agwu calls her work "God's Own…" One may also ask: What brings God into the genesis of mathematical story-telling? Encounter with Igbology reveals that God is presented as the author of the Igbo four-day week calendar. In effect, God is the root and origin of all knowledge, including mathematics.

It is interesting that Dr. Nkechi's story-telling begins with her childhood experience during the Nigeria-Biafra war and she was lucky to have survived the war by acts of providence. She happily credits her indomitable spirit to her Biafran war experience. This reviewer happens to have been a Biafran soldier and Biafran Marine during the war. I have carried out some mathematical projects with Dr. Nkechi and we are still working on some aspects of Igbo scientific culture. Dr. Nkechi's mathematical experience will be of great help in our Igbological agenda. Mathematics provides us with instruments of labor in our project on Igbo Cultural Symbols.

The future of all natural sciences depends on the appropriate application of symbols because no science can operate without the use of symbols.

Fada Jon Ofoegbu Ukaegbu, Ph.D.
Professor of Anthropology, Department of Social and Behavioral Sciences
Medgar Evers College, City University of New York

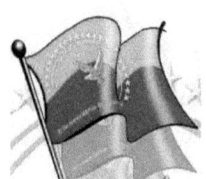

United Peoples Inclusive, Inc.
1601 Nostrand Avenue, Brooklyn, NY 11226
Tele. 718-284-4453; Email: gcr.lglg@yahoo.com

"We Come Together Just A Few, To Give Of Ourselves   Wisely, That All May Receive Sufficiently."

Experience being the ultimate teacher, had I read this autobiography of Dr. Nkechi Madonna Agwu before I became acquainted with her and her exceptional scholarship, incredible life-threatening ordeals and unrelenting will, I would have relegated this text to my fictional or fairytale book shelves. At five years of age, almost still a toddler, Dr. Agwu was absorbed in the 1967 Nigerian or Biafran Civil War. Civil Wars generally exact more carnage than that of wars between nations. This particular civil war claimed the lives of more than two million innocent men, women and children; and this silly display of pride or arrogance was avoidable.

Abruptly rooted from one of the most secured childhood homestead in all of Africa, five-year old Nkechi was on the run for her life. Homelessness, refugee camps, and all that are akin to these conditions ensued. Nonetheless, in just a few short decades, against all the odds, Dr. Agwu rose to worldwide leadership rank in the field of Mathematical Numerology fostering authentic story-telling in the lifestyle and rural culture of African women in particular. This text is a phenomenal test of will; it must be read. With an open mind, you will find it transformative to the vertex of your *Devine Destination*.

Austin Tuitt, *Elder T*
Founder/Director of United Peoples Inclusive

ANAWIM CLINICAL PASTORALTRAINING CENTER, "ACPTC", Inc.

143 W 117 Street, New York, NY 10026

Tel: 917-376-7109
Fax: (212) 864-9722
Email:acptclnc1@gmail.com

A Review

Professor Nkechi Nma Jacob Agwu has become very famous because of her 'Mathematical Genome in the tale of her Vertex-Edge Graphs and Numbers.' Having known Dr. Nkechi, namesake of my wife, for about 15 years through the American Association of University Women (AAUW) in New York City, I admire her positivity in approach to life, erudition, and complex simplicity.

Dr. Nkechi's life experience is important for many young girls and women who have not experienced suffering as she did from her very young age of 5 during the Biafra and Nigeria war. Young women should learn from her story the best way to show paternal love for fathers who, naturally and psychologically, love their daughters more than sons. Her great accomplishments are worth publishing to portray the real emancipation of women regardless of country of birth, language, color, or the condition they might find themselves.

In telling her story, Dr. Nkechi Nma Jacob is open and in touch with her real humanity as anyone else. This is to encourage young women to work hard rather than hide behind the vicissitudes of life. She did not shun talking about the ravages of sicknesses such as Diabetes, nor did she hide her sufferings, early widowhood and mortality of her loved ones during and after the senseless attack on Biafra (1967- 1970), where I ended up commanding the 20th Infantry Battalion under Division 12 of the Biafra Army.

Her mantra of "Do what you can with what you have everywhere (and whenever) you can" shows her indomitable spirit and determination to use every material object within her reach to teach and educate people 'Mathematically.' As a Minor in Calculus at Baruch College, CUNY, (1978), I appreciate her deep research into the History of Mathematics which is basically the beginning of everything.

God's Own: The Genesis of Mathematical Story-Telling

Finally her deep rooted faith in, and appreciation of, God, family, Church, country and even continents, is something everyone should emulate, especially after one emancipates from low to high life, from darkness to light, and from poverty to wealth. Professor Nkechi Nma Jacob has become a household name among young girls everywhere.

Dr. Emeka Nwigwe, Ph.D. (Venerable Chaplain)

(Psychoanalyst Intern)

NYPD Chaplain, 26th Precinct (Auxiliary Section)

Rector, Anglican Church of the Pentecost International, NYC: (website: angchurchpentintl.org)

Ven. Emeka C. Nwigwe, PhD.
Director/Training Supervisor
CPSP Board Certified Clinical
Chaplain/Pastoral Counselor

Glorious Miracle Embassy International
(Power House)
Umuahia Branch: Ehimiri Housing Estate by Zone 9 Police Office
Abuja Branch: Dakibiu by Jabi Upstay
Contact #: +234-8033911826

God's Own: A Genesis of Mathematical Story-Telling is a platform of life in general. The world itself was created by God's divine mathematical calculations. In the world today, no man or woman can function effectively in life without a calculative mentality, just like the saying, if you fail to plan, you are planning to fail. In other words, you were created by God to be mathematical in thinking.

This good book of the Lord says that which is in the Bible in Psalm 24:2. For God laid the earth's foundation on the seas and built it on the oceans depth. This is a great mathematical illustration of creating the universe, telling the true story of His creation. Through this autobiography, you will be inspired and provided with wisdom on how to handle issues related to this subject.

As a man of God, the very first day I heard about the title of this book and the formula in which the author, Professor Nkechi Agwu, uses to write this autobiography, I was inspired by the great wisdom God has given her to be a blessing to the world and to demonstrate His ability to move mountains in a person's life.

Prophet (Rev.) Emmanuel Angel
Founder and General Overseer of Glorious Miracle Embassy International
Author of *Moving Mountains: A Closer Walk With God*

About the Author

Dr. Nkechi Madonna Adeleine Agwu, aka Nma (Beautiful) Jacob is a tenured Professor of Mathematics and former Director of the Center for Excellence in Teaching, Learning and Scholarship (CETLS) at the Borough of Manhattan Community College (BMCC), City University of New York (CUNY). She was a President of the American Association of University Women, New York City Branch Incorporated.

Again, it is not by accident that there are seven Ndebele Dolls in the picture of Dr. Agwu above. As indicated in the review section of this book, the number seven symbolizes completeness and the creative timeline of God as reflected in the book of Genesis, chapters 1 and 2 in the Bible. It also symbolizes the miracle of exponential growth in the five loaves and two fishes that feed over 5000 men, women and children with 12 baskets of leftovers. The seven Ndebele dolls are birthing the miracle of exponential growth with all the people who will be touched by this book.

Dr. Agwu is nationally recognized for developing curriculum materials for innovative teaching as a recipient of several honors, including a Carnegie African Diaspora Fellowship (CADF), a CUNY Excellence Award, an American Mathematics Association of Two Year Colleges (AMATYC) INPUT Award and a New York City Literacy Assistance Center Mini-grant Award, *Effective Teaching Techniques Inherent in the Informal Educational System of the Igbo Culture*.

Dr. Agwu is a Project Kaleidoscope (PKAL) of the Association of American Colleges and Universities (AAC&U) Faculty for the 21st Century honoree, Class of 1997. She chairs the Social Advocacy Committee of Black Women for Black Girls Giving Circle (BWBG) and worked in partnership with Brotherhood Sistersol Incorporated on a Civic Fellowship Project with high school girls to develop a children's book on black girls and self-esteem based on the findings of the BWBG research report, *Black Girls in New York City: Untold Strength and Resilience*. She is a volunteer fire-fighter, a trained NYS community emergency response team (CERT) member, and a NYS Chaplain certified to practice world-wide through the Worldwide Association of Small Churches.

Dr. Agwu is author and editor of several scholarly publications, including the Writing Team Chair for the Chapter on Teaching for AMATYC's signature document *Beyond Crossroads*. Her ethno-mathematics research and co-curricular development CADF Project, *Culture and Women's Stories: A Framework for Capacity Building in Science, Technology, Engineering and Mathematics (STEM) Related Fields,* is based on harnessing the indigenous scientific and mathematical knowledge evident in our African cultures for functional education that facilitates creativity, innovation and entrepreneurship; mathematical that facilitates mathematical and cultural story-telling of successful African women in STEM using vertex-edge graphs, number patterns and geometric symbolism; that nurture the African girl child to consider STEM related careers; and that fosters development of curricular materials for teaching mathematics in low-income and rural areas using cultural, environmental and local resources.

Dr. Agwu is a mathematical story-teller of the Nigerian Women in Agricultural Research for Development (NiWARD). She writes poetry as a hobby. She is an artist with a passion for Ndebele Doll sculpturing of African women leaders in science, technology, engineering and mathematics (STEM) related fields. She is a 10th generation farmer from Umuapu, Agbakoli, Akoliufu, Alayi. She plants mathematical, cultural, and spiritual seeds in humans. She is God's Own (Nkechi), Nma (Beautiful) Jacob, and His faithful servant, created in His like image! According to His promise in Daniel 11:32, she shall perform exploits. This autobiography is just one of many to come.